The NEW Atheism

The NEW Atheism

Taking a Stand for Science and Reason

VICTOR J. STENGER

59 John Glenn Drive
Amherst, New York 14228–2119

Published 2009 by Prometheus Books

Inquiries should be addressed to
Prometheus Books
59 John Glenn Drive
Amherst, New York 14228–2119
VOICE: 716–691–0133, ext. 210
FAX: 716–691–0137
WWW.PROMETHEUSBOOKS.COM

13 12 11 10 09 5 4 3 2 1

Library of Congress Cataloging-in-Publication Data

Stenger, Victor J., 1935–
The new atheism : taking a stand for science and reason / Victor J. Stenger.
p. cm.
Includes bibliographical references and index.
ISBN 978–1–59102–751–5 (pbk. : alk. paper)
1. Atheism. 2. Religion. 3. Christianity and atheism. 4. Religion and science.
I. Title.

BL2747.3.S737 2009
211'.8—dc22

2009019129

Printed in the United States of America on acid-free paper.

Dedicated to

Paul Kurtz,

who has contributed more to the advance of science and reason
than any other of his generation.

CONTENTS

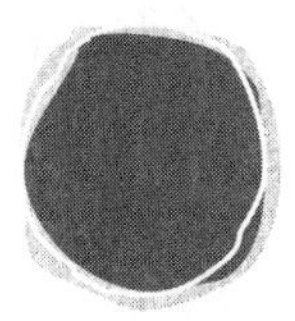

INGERSOLL'S VOW

Robert G. Ingersoll was a Peoria, Illinois, lawyer who became a famous orator in the period after the Civil War when oratory was a form of popular entertainment. Although he spoke on many subjects, he is remembered for his eloquent advocacy of free thought, humanism, and agnosticism.

When I became convinced that the Universe is natural—that all the ghosts and gods are myths, there entered into my brain, into my soul, into every drop of my blood, the sense, the feeling, the joy of freedom. The walls of my prison crumbled and fell, the dungeon was flooded with light, and all the bolts, and bars, and manacles became dust. I was no longer a servant, a serf, or a slave. There was for me no master in all the wide world—not even in infinite space. I was free—free to think, to express my thoughts—free to live to my own ideal—free to live for myself and those I loved—free to use all my faculties, all my senses—free to spread imagination's wings—free to investigate, to guess and dream and hope—free to judge and determine for myself—free to reject all ignorant and cruel creeds, all the "inspired" books that savages have produced, and all the barbarous legends of the past—free from popes and priests—free from all the "called" and "set apart"—free from sanctified mistakes and holy lies—free from the fear of eternal pain—free from the winged monsters of the night—free from devils, ghosts, and gods. For the first time I was free. There were no prohibited places in all the realms of thought—no air, no space, where fancy could not spread her painted wings—no chains for my limbs—no lashes for my back—no fires for my flesh—no master's frown or threat—no following another's steps—no need to bow, or cringe, or crawl, or utter lying words. I was free. I stood erect and fearlessly, joyously, faced all worlds.

And then my heart was filled with gratitude, with thankfulness, and went out in love to all the heroes, the thinkers who gave their lives for the liberty of hand and brain—for the freedom of labor and thought—to those who fell in the fierce fields of war, to those who died in dungeons bound with chains—to those who proudly mounted scaffold's stairs—to those whose bones were crushed, whose flesh was scarred and torn—to those by fire consumed—to all the wise, the good, the brave of every land, whose thoughts and deeds have given freedom to the sons of men. And then I vowed to grasp the torch that they had held, and hold it high, that light might conquer darkness still.

—Robert G. Ingersoll (1833–1899)

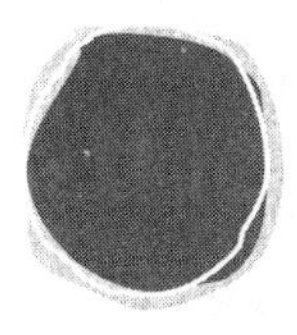

PREFACE

> **There are, after all, atheists who say they wish the fable were true but are unable to suspend the requisite disbelief, or who have relinquished belief only with regret. To this I reply: who wishes that there was a permanent, unalterable celestial despotism that subjected us to continual surveillance and could convict us of thought-crime, and who regarded us as its private property even after we died? How happy we ought to be, at the reflection that there exists not a shred of respectable evidence to support such a horrible hypothesis.**
>
> —Christopher Hitchens[1]

In 2004 Sam Harris published *The End of Faith: Religion, Terror, and the Future of Reason.* This marked the first of a series of six best-selling books that took a harder line against religion than had been the custom among secularists. This movement has been termed *New Atheism.*

Motivated primarily by the events of September 11, 2001, which he laid directly at the feet of the religion of Islam, Harris did not leave Christianity or Judaism off the hook. Nor did he pardon religious moderates:

> One of the central themes of this book . . . is that religious moderates are the bearers of a terrible dogma: they imagine that the path to peace will be paved once each of us has learned to respect the unjustified beliefs of others. I hope to show that the very ideal of religious tolerance—born of the notion that every human being should be free to believe whatever he wants about God—is one of the principle forces driving us toward the abyss.[2]

Harris followed in 2006 with *Letter to a Christian Nation*, which responded to the many messages he had received from Christians objecting strongly to his criticism of their faith, and of faith in general.

The same year, the famed biologist and author Richard Dawkins published *The God Delusion*, which stood on the *New York Times* Best Seller List for fifty-one weeks. Dawkins says he intended his book to be a series of "consciousness-raising" messages aimed at people who have been brought up in some religion or other who want to be free of it for one reason or another. Dawkins's first message: "You can be an atheist who is happy, balanced, moral, and intellectually fulfilled."[3]

Also in 2006, philosopher Daniel C. Dennett produced *Breaking the Spell: Religion as a Natural Phenomenon*, which made a plea for religion to be examined scientifically in order to better understand where religion came from, why people believe in God so fervently, and why religion is such a potent force in society.[4] Although the author insisted he was not taking a stand against religion, his previous reputation as an ardent proponent of atheism marked him as a new atheist.

In 2007 my own contribution, *God: The Failed Hypothesis—How Science Shows That God Does Not Exist* made the *New York Times* Best Seller List in March of that year.[5] With endorsements by Richard Dawkins and Sam Harris, my book fit right in with the genre of new atheist literature. In *God: The Failed Hypothesis* I made a unique argument. Scientists and others had written scores of books showing that there is no evidence for the existence of God. They were always countered by the truism "Absence of evidence is not evidence of absence." But I pointed out that absence of evidence *that should be there* is valid evidence of absence. I demonstrated that the absence of evidence that should be there is now sufficient to conclude beyond a reasonable doubt that the God worshipped by Jews, Christians, and Muslims does not exist.

In June of that year, *God Is Not Great: How Religion Poisons Everything* by journalist and literary critic Christopher Hitchens debuted in first place on the *New York Times* Best Seller List.[6]

Hitchens provided further evidence and arguments for the malign influence of religion in the world. He thus became recognized as a charter member of the New Atheism. Hitchens graciously wrote a foreword for the paperback edition of *God: The Failed Hypothesis* that came out in 2008.

As presented by these authors, the New Atheism breaks new ground that is not covered in other writings on atheism, both earlier and current. Most of these other writings are excellent and not in significant disagreement with the new approach on all but a few issues. Their emphases tend to be either philosophical, as, for example, with George H. Smith's classic *Atheism: The Case against God*,[7] or personal, as, for example, with Dan Barker's excellent *Godless*.[8]

The new atheists write mainly from a scientific perspective. Dawkins and I are science PhDs who have spent years doing scientific research, Dawkins in biology and I in physics and astronomy. Harris has a degree in philosophy and is working on a PhD in neuroscience. Dennett is a philosopher of science who has written almost exclusively on scientific topics. While Hitchens is not a scientist, his approach to religion is emphatically empirical. He has observed the lethal effects of religion firsthand during his travels as a foreign correspondent. All of us have been criticized for not paying enough attention to modern theology. We are more interested in observing the world and taking our lessons from those observations than debating finer points of scriptures that are probably no more than fables to begin with.

In this book I review and expand upon the principles of New Atheism. Not all nonbelievers—atheists, agnostics, humanists, or freethinkers—have been happy with the approach taken by the new atheists, especially our unwillingness to take a benign view of moderate religion. They would like to maintain good relations with the religious community, especially with regard to the public acceptance of science. They worry that government funding for science might be put at risk and the teaching of evolution in the schools compromised.

While new atheists sympathize with these concerns, we do not consider them as serious as the even greater dangers imposed by the irrational thinking associated with religion. Most people recognize the value and necessity of scientific research, especially medical, and are not likely to change that to spite a few loudmouthed atheists. And a child is not going to be scarred for life by hearing the dreaded word "creationism" in the classroom. We are not saying this issue isn't important. No one alive is a greater proponent of evolution than Richard Dawkins, and he recognizes what he calls the "superficial appeal" of all evolutionists working together to fight creationism. But he takes the position of geneticist Jerry Coyne that "the *real* war is between rationalism and superstition."[9] Another place where we differ from mainstream scientists is in the common but we think mistaken notion that science and religion are, in the words of the late, famed paleontologist Stephen Jay Gould, two "non-overlapping magisteria (NOMA)."[10] We agree with most reviewers that Gould's interpretation is incorrect and amounts to a redefinition of religion as "moral philosophy." Religions make statements about all kinds of phenomena that are legitimate parts of science, such as the origin of the universe and evolution of life. Even the principles of morality are subject to scientific investigation since they involve observable human behavior. Furthermore, we do not see morality as god-given but rather the result of humanity's own social development.

We strongly disagree with the National Academy of Sciences, and many scientists, that science has nothing to say about God or the supernatural. The gods most people worship purportedly play an active role in the universe and in human lives. This activity should result in observable phenomena, and it is observable phenomena that form the very basis of scientific investigation.

Numerous books vehemently protesting the New Atheism have been written by theists, almost all Christians. Their arrows have been mainly directed at Dawkins and Harris. As I will show in detail, their criticisms are misguided and their arguments easily countered. I examined a large sample of these books and did not find a single new argument for the existence of God or for the value of religion that had not already been addressed in atheist literature. So, I will be going over some familiar ground. However, I do significantly update some of the arguments against God, give a detailed account of naturalism, and follow up on a proposal by Harris in *The End of Faith* that we pay some attention to the ancient wisdom of the East.

Many theist authors present several arguments from physics and cosmology that have lain mostly unanswered in popular books. Where these arguments do come up in the writings of other physicists, they are usually dismissed therein as nonsense with little explanation. In several books now I have attempted to take these issues seriously and tried to come up with adequate natural explanations. I continue this effort here, with some new reasons to reject the claims that the big bang supports a supernatural creation and that the physical parameters of the universe are "fine-tuned" for our kind of life.

In this regard, I must repeat what I say in all my books about interpreting evidence. A good scientist does not approach the analysis of evidence with a mind shut like a trap door against unwelcome conclusions. If and when anyone finds evidence for the existence of God, gods, or the supernatural that stands up under the same stringent tests that are applied in science to any claimed new phenomenon, with no plausible natural explanation, then honest atheists will have to become at least tentative believers. This is not impossible, as I will show.

Perhaps the most unique position of New Atheism is that faith, which is belief without supportive evidence, should not be given the respect, even deference, it obtains in modern society. Faith is always foolish and leads to many of the evils of society. The theist argument that science and reason are also based in faith is specious. Faith is belief in the absence of supportive evidence. Science is belief in the presence of supportive evidence. And reason is just the procedure by which humans ensure that their conclusions are consistent with the theory that produced them and with the data that test those conclusions.

I have followed the other new atheists in cataloging a sample of the atrocities associated with religion, both in the Bible and in history. The common theist response that atheists in the twentieth century killed more humans than theists over the ages is not supported by the facts. Furthermore, nonbelievers do not kill in the name of atheism the way believers kill in the name of God. We will study the example of Mormonism to see how easy it is for people to commit the most horrible acts when they are convinced they are doing so under orders from God.

I will also show how Judaism, Christianity, and Islam cannot account satisfactorily for the suffering in the world and certainly have done little to alleviate it. We will discover that 2,500 years ago the Buddha and other sages of the East provided a way to endure suffering that is still applicable today. By meditation, ritual, or other means we rid ourselves of the self-centeredness that so dominates our lives.

While Judaism, Christianity, and Islam have produced similar teachings, they are not practiced by most adherents. In fact, Christianity and Islam are the most popular religions in the world because of their appeal to our most selfish instincts with the promise of eternal life. The followers of the Hindu and Buddhist religions also often ignore the best wisdom in their philosophical traditions, pursuing egocentric desires through supernatural hopes.

I have followed Harris's suggestion that we consider the teachings of the sages of the East and concluded that these teachings can be stripped of any supernatural baggage that is implied by referring to them as "spiritual" or "mystical" and still have the same force. An atheist can meditate without any mumbo-jumbo and arrive at the state of mind of a Zen Buddhist who reaches the understanding that the ego is artificial. As we will see, brain imaging is beginning to provide preliminary evidence for how this takes place purely naturally by the shutting down of the part of the brain where ego seems to be located.

Atheism cannot compete with any of the supernatural religions that disingenuously promise eternal life.[11] However, all the evidence points to a purely material universe, including the bodies and brains of humans, without the need to introduce soul, spirit, or anything immaterial. But, repeating the quote from Richard Dawkins given above, "You can be an atheist who is happy, balanced, moral, and intellectually fulfilled." Once we are free of the fetters of faith we can look forward to a life under our own control.

Basically, then, this book debunks many of the myths about religion and atheism that are held not only by believers but by many nonbelievers as well. One final myth is that religion is growing in the world and secularism is dying. As we will see, the facts tell a different story. There are from one to two billion nonbelievers on this planet, depending on how you count the Chinese. This makes nonbelief either the second or third "belief system," exceeded only by Christianity and possibly Islam. And, it is the fastest growing.

This growth is mainly confined to "postindustrial" nations where at least two-thirds of the gross domestic product comes from services. Belief in agricultural and industrial (manufacturing) societies has largely remained flat over a century. In general, the more wealthy a nation, the less likely it is to be religious. Ireland, Italy, Canada, Finland, and especially the

United States are anomalously more religious than expected for their wealth. On the other hand, Denmark, the Netherlands, and France are anomalously low. These anomalies cannot be attributed to a single cause but are probably the result of a mix of factors, such as history and economic inequality.

Most believers have been brainwashed into thinking that religion is necessary for happiness and contentment. This flies in the face of the fact that the happiest, healthiest, most content societies are the least religious. The new atheists are not trying to take away the comfort of faith. We are trying to show that life is much more comfortable without it.

The new atheists are committed to helping accelerate the trend away from religion that is already occurring in certain parts of the world. We ask other atheists and agnostics to join us in taking a harder line against the follies and atrocities of religion produced by its irrational thinking. Not only will a more secular world improve our security by making wars more unlikely, it will allow science and reason to once more help guide government policies, especially in the United States after eight years or more of being ignored in favor of "faith-based initiatives." We see this as the only road to survival.

NOTES

1. Christopher Hitchens, ed., *The Portable Atheist: Essential Readings for the Nonbeliever* (Philadelphia: Da Capo Press, 2007), p. xxii.

2. Sam Harris, *The End of Faith: Religion, Terror, and the Future of Reason* (New York: Norton, 2004), pp. 14–15; audiobook available from Blackstone Audio, Inc. (2006).

3. Richard Dawkins, *The Selfish Gene* (New York: Oxford University Press, 1976), p. 1.

4. Daniel C. Dennett, *Breaking the Spell: Religion as a Natural Phenomenon* (New York: Viking, 2006).

5. Victor J. Stenger, *God: The Failed Hypothesis—How Science Shows That God Does Not Exist* (Amherst, NY: Prometheus Books, 2008).

6. Christopher Hitchens, *God Is Not Great: How Religion Poisons Everything* (New York: Twelve Books, 2007).

7. George H. Smith, *Atheism: The Case against God* (Amherst, NY: Prometheus Books, 1989).

8. Dan Barker, *Godless: How an Evangelical Preacher Became One of America's Leading Atheists* (Berkeley, CA: Ulysses Press, 2008).

9. Richard Dawkins, *The God Delusion* (Boston: Houghton Mifflin, 2008), p. 67.

10. Stephen Jay Gould, *Rocks of Ages: Science and Religion in the Fullness of Life*, Library of Contemporary Thought (New York: Ballantine, 1999).

11. In the definition of atheism used in this book, I do not include Buddhists or any other nontheists who still hold to supernatural beliefs. This may be a bit contradictory, however, it corresponds more closely to common usage and so should lead to less misunderstanding.

ATHEISM ON THE OFFENSIVE

Religion is a disease.

Heraclitus (c. 535—475 BCE)

THE GODS PEOPLE WORSHIP

Many nonbelievers will tell you that they are not "atheists" but "agnostics." Although they see no evidence for God, they reason that we can never know whether or not God exists. After all, they say, "Absence of evidence is not evidence of absence." How can we possibly know that a god does not exist who is hidden from us so that we have no proof or evidence either way?

For example, an impersonal god may have created the universe and then left it alone to follow its own path, governed by the natural laws that it built into the universe, perhaps along with a large element of chance. This doctrine is called *deism.* As I discussed in my previous book, *Quantum Gods*, such a god who, in Einstein's famous words, "plays dice with the universe" might be impossible to detect.[1]

Einstein often used the word "god" in his philosophical statements, but made it clear that he did not believe in a personal god. When pressed he said he believed in the "god of Spinoza," who is basically a metaphor for the order and structure of the universe. This doctrine is called *pantheism.* A related doctrine is *panentheism,* in which god includes the physical universe and whatever else there is. Pantheism simply gives the name "god" to the sum of reality and is empty of any religious content that can be used to guide people's lives or ritual that can be used to provide inspiration. The god of the Jewish kabbalistic tradition is panentheistic and religiously significant. However, neither of these views plays an important role in modern religious thinking.

Over the ages the most common form of god worship has been *polytheism,* the worship of many gods, usually representing various objects in nature from the sun and moon to animals and humans. Today, Hinduism is the only truly polytheistic religion practiced by a large number of people, although technically it is called *henotheistic* in that there is only one absolute reality, Brahman, and the various gods of the Hindu pantheon are merely representations (avatars) of this one reality.[2] Muslims consider the Christian Trinity polytheistic, and Catholics pray to a whole constellation of saints who in another time would be called gods. In his documentary film *Religulous,* Bill Maher hilariously interviews a "senior Vatican priest" standing outside St. Peter's who tells him that Jesus Christ is only sixth on the list of heavenly personages that Italian Christians pray to.[3]

But, such differences aside, the god that more people currently worship than any other is some variation on the personal god of the Jewish Torah, Christian Old Testament, and Muslim Qur'an. He is called YHWH in the Torah and Old Testament, and Allah in the Qur'an.

I will, for simplicity, often refer to this god as the "Abrahamic God," or just "God," which should be understood to refer to the god of the three religions that trace their origins to the patriarch Abraham, a probably mythological figure who supposedly lived four thousand years ago.[4] If Abraham existed at all, he would have been polytheistic, as were all the peoples in Canaan at the time. Judaism did not become monotheistic until after the return of the Jews from exile in Babylon circa 530 BCE.

THE ATHEIST WORLDVIEW

The term *theism* usually refers to the belief in a personal god or gods such as found in Judaism, Christianity, Islam, and Hinduism. Technically then, an atheist is someone who does not believe in the gods of these religions. If atheists were defined as people who are not theists, they would include deists, Buddhists, and others generally identified as "spiritualists." Deists believe in a god who created the universe but then leaves it alone to carry on according to the laws he has laid out, rarely if ever stepping in to change the course of events and not paying much if any attention to individual lives. Buddhists (though not all) do not believe in gods but have some sense of a supernatural. Spiritualists believe in an unknown higher power, a so-called universal spirit, or simply have a vague notion that "something must be out there." Let us stick to the common usage of atheist to mean someone who believes that no gods exist, including the deist god, the universal spirit, or any other vague possibility that agnostics prefer to leave open. This includes Buddhists. That is, there are theists, deists, Buddhists, spiritualists, agnostics, and atheists. More simply, there are believers and nonbelievers, or supernaturalists and naturalists.

Atheists view science as the best means humanity has yet come up with for understanding the world. Not all scientists are atheists. However, few of those scientists who are believers assume any role for the supernatural in explaining the phenomena that reach our senses. Rather they compartmentalize their thinking into scientific and religious. They enter one compartment when they go to church on Sunday, one that has been swept clean of the critical thinking and devil's advocacy that came with their training. Monday morning they return to work and enter the other compartment, where God never enters the equations.

As you often hear, "science is provisional." So both believing and nonbelieving scientists must keep open the possibility that their concepts may change should the data warrant it. Science just makes models anyway and does not require nor does it use any metaphysics. As far as science can tell, the universe is matter and nothing more. If at some future date, scientists find they need something other than matter in their models and they can still describe that substance mathematically, this new stuff would not be supernatural. In that case a distinction between believer and atheist would still remain. For now, let's keep things simple: the atheist believes that we need not include anything beyond matter (to be defined precisely later) to describe the universe and its contents. Everybody else thinks we need something else.

The atheist view is not what some believers derogatorily call "scientism," the view that science is the only source of knowledge. Atheists appreciate the beauty of art, music, and poetry as much as believers, along with the joys of love, friendship, parenthood, and other human relationships. We love life even more than the believer, because that is all we have. We only insist that when anyone makes a claim about the world of our senses, that science and reason be allowed to test that claim.

That may sound noncontroversial, but one of the major complaints that the new atheists have about society today, especially in the United States, is that religion is given a special dispensation from the requirement of rationality that is applied in all other forms of human discourse. We are pilloried for hurting people's deepest feelings when we cast doubt that somebody born of a virgin rose from the dead or when we question that some book contains all the truths that anyone need ever know. We not only regard such beliefs as wrong, we see them as immoral and dangerous to the future of society.

Philosopher Michael Martin, who has written authoritatively on atheism, distinguishes between two types, *negative atheism* that simply identifies nonbelief, including agnosticism, and *positive atheism*, where the nonbeliever "rejects the theistic God and with it the belief in an afterlife, in a cosmic destiny, in an immortal soul, in the revealed nature of the Bible and Qur'an, and in a religious foundation for morality.[5]

A breakdown of world belief is given in Table 1.1.

Table 1.1. Major religions of the world (partial list) ranked by number of adherents.[6]

Religion	*Number*	*Percentage*
Christianity	2.1 billion	33
Islam	1.5 billion	21
Secular/Nonreligious/Agnostic/Atheist	1.1 billion	16
Hinduism	900 million	14
Chinese traditional	394 million	6
Buddhism	376 million	6
Judaism	14 million	0.22

These estimates were obtained from the Web site of a Christian organization, adherents.com, that based them on *Encyclopedia Britannica* and *World Christian Encyclopedia* reports.[7] The numbers for each faith are admitted to tend toward the high end since they count people with a minimal level of self-identification.

We see that there are over a billion nonbelievers in the world, more than Hindus, making it the third-largest "belief" system. This may be a vast underestimate that has not counted China accurately. I have seen estimates that there are as many as a billion nonbelieving Chinese alone. But while they may not believe in gods, most Chinese maintain traditional supernatural beliefs such as feng shui and fortune telling.

The table does not break down the nonbelievers. Martin quotes an estimate from the 2002 *New York Times Almanac* that 4 percent of the world's population are professed atheists. With an estimated world population at this writing of 6.7 billion, this gives 268 million current admitted atheists, while there are no doubt many more who keep their nonbelief to themselves.

Despite the apparent fact that the people of the United States are exceptionally religious, at least in terms of professed supernatural beliefs of one type or another, the national figures are not much different from those worldwide when surveys probe more deeply than simply asking the subject to state a religious preference. While only 1.6 percent, or about 4.8 million Americans admit they are atheists, this is undoubtedly an underestimate because the term "atheist" has a more negative connotation in America than in other developed countries. In fact, the 2008 American Religious Identification Survey estimates from stated beliefs that 12 percent or 36 million Americans are atheists.[8] Atheists fall below homosexuals in the esteem of a majority of Americans. For example, in the November 2008 US elections, North Carolina senator Elizabeth Dole, wife of 1996 Republican presidential candidate Robert Dole, put out a TV ad that her Democratic opponent, Kay Hagan, accepted money from an atheist rights group.[9] A Dole commercial showed pictures Hagen with a woman's voiceover saying, "God does not exist." Hagan angrily filed a lawsuit for defamation. In America, apparently, calling a person an atheist is as bad as calling her a child molester. Hagan handily defeated Dole, while the lawsuit is still pending as of this writing.

IS AMERICA A DEIST NATION?

Although a majority of Americans call themselves "Christians," a study of their actual beliefs made by Baylor University in 2005 indicates that many people who think of themselves as Christians actually disagree with basic Christian teachings. The survey found that 44 percent of Americans do not believe in a god who plays an important role in the universe or their personal lives.[10] That makes them *deists*, not theists, although most probably would not accept the designation. But the fact is that only 54 percent of Americans are true Christians who believe in a God who significantly acts in the universe and in human lives. Certainly you are not a Christian if you do not believe in a God who consciously acts in the world.

So more than a billion people on Earth do not believe in the Abrahamic God or any other specific god, outnumbering Hindus and Jews, and almost matching the number of Muslims. If some fraction of the estimated Chinese nonbelievers is counted, then there are more nonbelievers than Muslims and only Christians exceed nonbelievers.

ATHEISM IN SOCIETY

Despite the large numbers of nonbelievers, atheism has not represented as major a force in world or national affairs as many much smaller groups such as Jews or homosexuals. While communism is officially atheist, a fact Christians gleefully point to as an example of what happens when atheism holds sway, communist philosophy is based on economics, not metaphysics. The political and economic philosophies of communism do not follow from the hypothesis of nonbelief. Indeed, with its dogmatic policies and authoritarianism, communism more closely resembles a godless religion than secular atheism. The only political views of atheists that are closely linked to their beliefs are that church and state should be separate and that decisions should be based on reason rather than revelation.

A number of organizations in the United States and elsewhere promote atheist, freethinking, and secular humanist viewpoints. The organization with the largest membership of freethinkers is the Freedom from Religion Foundation based in Madison, Wisconsin. Its copresidents are Dan Barker and his wife, Annie Laurie Gaylor. For nineteen years Barker was an evangelical preacher and highly successful composer and performer of Christian music. Gradually, without any atheist proselytizing, he came to doubt and finally disbelieve in Christianity and all religion for that matter. Barker tells his story in his 2008 book, *Godless: How an Evangelical Preacher Became One of America's Leading Atheists.*[11]

Most prominent atheists contribute regularly to *Free Inquiry* magazine, published by the Center for Inquiry located in Amherst, New York. CFI is a think tank for atheism and secular humanism.[12] It also coordinates the local efforts of several other centers along with a large number of community and student groups worldwide.

Currently, the net memberships of FFRF, CFI, and other organizations are comparatively small and hardly the threat that Christian leaders would like Christians to think they are.[13] In the past, nonbeliever organizations have not been able to recruit a major fraction of those who hold similar views on God and religion and stir them up into political action.

I think it is fair to say that, at least until recently, the typical nonbeliever simply did not give religion much thought and lived her life as if it were the only one she had—which happens to be the case. Nonbeliever publications and other media outlets have had little influence when compared to the vast industry of religious books, magazines, TV and radio stations, and Internet sites. One reason why America seems to be more religious than other developed countries could be the vast Christian propaganda machine, which is unmatched anywhere else. In the past the subjects of atheism and secularism have received scant attention in the general media. Until recently, books on atheism have not sold well, although some atheist authors, notably Richard Dawkins and Daniel C. Dennett, have written best sellers on the closely related subject of evolution.

NEW ATHEISM STEPS IN

This all changed dramatically in 2004 with the first in a series of best sellers by authors who preached a more militant, in-your-face kind of atheism that had not been seen before, except with the abrasive and unpopular Madalyn Murray O'Hair. This phenomenon has been termed *New Atheism* and it may augur well for the future of nonbelief. New Atheism seems to be a growing phenomenon in the United States and has attracted much media attention.[14] The atheist and freethinker groups mentioned above are experiencing increasing membership, especially on college campuses.

As you might expect, this phenomenon has driven Christian apologists to distraction. A whole raft of books has been published in response to the atheist best sellers, largely from Christian publishing houses, of which, as I said, there are many. Most of these anti-atheist screeds are marked by shoddy scholarship such as incomplete references, inaccurate quotations, and misrepresentations of atheist views. I will give specific examples in the course of this book. None of the anti-atheist books have sold anywhere near as well as the atheist books they challenge. In what follows I will quote from some of these and other commentaries as I summarize the arguments made in the new atheist literature. This first brush should give a good taste of the conflict and demonstrate why any reconciliation is unlikely in the near future. In the following chapters we will discuss the issues in greater detail and expand upon what I see as the message of New Atheism.

Here's how the well-known conservative author and political commentator Dinesh D'Souza describes the new atheists in his 2007 book, *What's So Great about Christianity?*

> The atheists no longer want to be tolerated. They want to monopolize the public square and to expel Christians from it. They want political questions like abortion to be divorced from religious and moral claims. They want to control school curricula so they can promote a secular ideology and undermine Christianity. They want to discredit the factual claims of religion, and they want to convince the rest of society that Christianity is not only mistaken but evil. They blame religion for the crimes of history and for the ongoing conflicts in the world today. In short, they want to make religion—and especially the Christian religion—disappear from the face of the earth.[15]

In this introduction D'Souza does not provide evidence for his assertions by quoting from the new atheist literature. We certainly do not want to "expel Christians" from the public square or "control school curricula," but we would not be unhappy with some of the other outcomes. As we will see later, when he does quote new atheist writers (including myself), D'Souza does so misleadingly and, in at least one important case, falsely.

In *The New Atheist Crusaders and Their Unholy Grail: The Misguided Quest to Destroy Your Faith*, Christian writer Becky Garrison, senior contributing editor of the religious satire magazine the *Wittenburg Door*, says,

> These new atheists aren't resurrecting the old atheist argument that belief in God is wrong. Rather, they're advocating that belief in God is dangerous and destructive. Furthermore, while old-school atheists came to the conclusion that God doesn't exist after some angst-ridden anxiety and serious soul-searching, this current crop of anti-God guys giggle like schoolgirls over their naughty refusal to kowtow to society and buy into this God biz.[16]

In *God and the New Atheism* the reputable theologian John Haught agrees with Garrison. As old-school atheists he mentions Friedrich Nietzsche, Albert Camus, and Jean-Paul Sartre as facing up to what the absence of God should really mean—the "disorienting wilderness of nihilism."[17] This must be the "angst-ridden anxiety and serious soul-searching" that Garrison is talking about.

Haught tells us, "Before settling into a truly atheistic worldview, you would have the experience the Nietzschean Madman's sensation of straying through 'infinite nothingness.'"

In short, you have to be willing to risk madness to be a true atheist. The new atheists, not being madmen, do not qualify.

I don't see how lifting the burdens of dreadful superstitions such as the fear of hell and replacing them with a coolly reasoned understanding of the universe should result in "angst-ridden anxiety." I have no trouble thinking about "infinite nothingness" and, in fact, I can represent it mathematically.[18] The madmen are those theists who hear the voice of God telling them to blow themselves up along with as many innocents as possible.

THE END OF FAITH?

The new atheist movement burst into existence in 2004 with the publication of *The End of Faith* by a young neuroscience graduate student with a degree in philosophy, Sam Harris.[19] Harris cuts straight to the heart of religious belief and identifies blind faith as the source of much of the unreason in the world and a prime contributor to the terrorism and fanaticism we have experienced in recent years, in particular September 11, 2001. He disputes the common claims that faith is an essential component of human life and that other sources than faith are responsible for the horrifying acts of fanatical believers:

> Two myths now keep faith beyond the fray of rational criticism, and they seem of foster religious extremism and religious moderation equally: (1) most of us believe that there are good things that people get from religious faith (e.g. strong communities, ethical behavior, spiritual experience) that cannot be had elsewhere; (2) many of us also believe that the terrible things that are sometimes done in the name of religion are the products not of *faith* per se but of our baser nature—forces like greed, hatred, and fear—for which religious beliefs are themselves the best (or even the only) remedy.[20]

Harris comes down hard on those such as linguistics professor Noam Chomsky who regard America as a terrorist nation that brought the September 11 attacks on itself.[21]

Fareed Zakaria, the editor of *Newsweek International* and columnist for the *Washington Post*, has attributed the turmoil in the Middle East to the failure of political institutions: "If there is one great cause of the rise of Islamic fundamentalism, it is the total failure of political institutions in the Arab world."[22] Harris responds, "Perhaps. But 'the rise of Islamic fundamentalism' is only a problem because *the fundamentals of Islam* are a problem."[23]

While addressing Islamic terrorism in some detail, Harris hardly leaves Christianity off the hook. Unlike most of the secular community, Harris does not give moderate and liberal religion, Muslim and Christian, a free ride. He asserts that by refusing to speak out against religious extremists and by holding to irrational beliefs of their own and insisting on the revealed nature of their scriptures, moderate Muslims and Christians provide aid and comfort to those who wish to force medieval ideas on the rest of society.

Quoting Deuteronomy 13:6–10, where God commands you to kill anyone, including your father or mother, if he or she suggests serving other gods, Harris notes that moderation in religion has nothing underwriting it but the "unacknowledged neglect of the letter of divine law."[24] He continues: "The only reason anyone is 'moderate' in matters of faith these days is that he has assimilated some of the fruits of the last two thousand years of human thought (democratic politics, scientific advancement on every front, concern for human rights, and end to cultural and geographic isolations, etc.)."[25]

In November 2006, Harris and Dawkins attended a meeting called Beyond Belief at the Salk Institute in San Diego. The videos of the entire conference were still on the Web as of this writing over two years later and very much worth watching, along with subsequent meetings.[26] Top scholars in science also attended and most of those who spoke were atheists.

I was somewhat taken aback by the benign view of religion presented by the atheistic scientists other than Harris and Dawkins. Nobel laureate physicist Steven Weinberg, an avowed atheist, has made many quotable statements against religion, such as: "The more the universe seems comprehensible, the more it also seems pointless."[27] And, "Religion is an insult to human dignity. With or without it you would have good people doing good things and evil people doing evil things. But for good people to do evil things, that takes religion."[28]

Yet, at the meeting he said he was more "sanguine" about religion than Harris. When he meets people socially where he lives in Texas, he finds that, when the subject of religion comes up, most believers do not hold to many of the extreme official views of the churches they attend. For example, few Christians think that a non-Christian who lives a good life cannot go to heaven. However, ask their preachers or theologians and they will tell you otherwise. Of course Weinberg's circle of acquaintances around the University of Texas at Austin is not likely to be representative of Americans elsewhere.

Other atheist speakers came down hard on Harris and Dawkins, arguing that their approach will not earn any converts to atheism and asking what right do atheists have to deny believers the comforts of faith. Harris and Dawkins tried valiantly to point out the dangers of continuing this reliance on dogma rather than reason and science as the world heads for a period of increasing turmoil. They make the case very eloquently in their speeches and books, which their detractors do not seem to have read very carefully. These detractors seemed totally oblivious, for example, to the fact that since Ronald Reagan our presidents have taken advice from preachers who view events in the Middle East as a precursor to the Second Coming of Jesus, when the world will come to an end. As we will see in the next chapter, millions of American Christians anticipate this occurrence.

Harris suggests an alternative to the irrationality of the Abrahamic religions that is not simply scientific materialism. He regards Judaism, Christianity, and Islam as hindering the development in the West, and in Islamic countries, of the real "spiritual" progress made in Buddhism and other Eastern religions over thousands of years. I think most new atheists would agree with me that we should avoid using the term "spiritual" to refer to phenomena such as love, compassion, and selflessness. We see no reason why these cannot result from purely material forces, and "spiritual" to most people implies something supernatural.

However, I have made my own independent study of ancient Eastern philosophy, which I will review in a later chapter. I find that when stripped of any implication of supernaturalism I agree with Harris that Eastern philosophers uncovered some unique insights into humanity and the human mind that were lacking in the West. I propose that the teachings of the ancient sages of the East constitute a "Way of Nature" that provides atheists and materialists with a viable path to peace and happiness. The sages' teachings are marked by selflessness and calm acceptance of the nothingness after death. Although a handful of mystics in Judaism, Christianity, and Islam preached a similar inward-looking selflessness, it was closely tied to the supernatural, whereas the Eastern methods did not require any supernatural element. The mass-scale religions of the West and the East ignored this ancient wisdom, replacing it with the extreme self-centeredness associated with the absurd promise of eternal life.

Continuing with the critics, Becky Garrison asserts, "I can attest that one can be a practicing Christian and still have some serious issues with the institutional church. Heck, I've been ranting against church crud since 1994, a good ten years before Harris hacked out *The End of Faith*."[29]

If there's anything I hate, it's an unholier-than-thou attitude.

In 2006 Harris came out with a small book called *Letter to a Christian Nation* that also made the best seller lists.[30] Here he responded to the thousands of letters he received after his first book telling him how wrong he was in his criticisms of both extremist and moderate religiosity. The most militant messages were from Christians, so Harris decided to reply specifically to them. As he explains on the first page,

> Christians generally imagine that no faith imparts the virtues of love and forgiveness more effectively than their own. The truth is that many who

> claim to be transformed by Christ's love are deeply, even murderously, intolerant of criticism. While we may want to ascribe this to human nature, it is clear that such hatred draws considerable support from the Bible. How do I know this? The most disturbed of my correspondents always cite chapter and verse.[31]

Harris has been severely criticized for his position on certain political issues such as torture and preemptive nuclear war. I will not discuss these in this book because they have nothing to do with atheism. Harris maintains a Web page, "Response to Criticism," where he clarifies and further explains his positions.[32]

THE GOD DELUSION

Two other best sellers also appeared in 2006 by authors far better known than Sam Harris: Oxford biologist Richard Dawkins and Tufts University philosophy professor Daniel C. Dennett.

Dawkins gained almost instant fame as a young researcher when in 1976 he published *The Selfish Gene*, which promoted the idea that evolution operates on genes rather than individual organisms.[33] That is, it is not the organism that seeks to survive but the genetic information that is collected in what is termed a gene.[34]

Evolutionary models are very difficult to verify or falsify and so often remain contentious within the field for years. Creationists often take misplaced comfort from this. No dispute among experts in the field can be found on the basic correctness of the Darwin-Wallace scheme of evolution by natural selection. Arguing over detailed mechanisms is the everyday *modus vivendi* of science.

However, Dawkins's selfish genes (based on earlier work) offered a tentative explanation for altruism since in the selfish gene scheme the evolutionary drive is the survival of genetic information. Thus the sacrifice of an individual, a common observation in nature, can increase the odds of the survival of a gene.

Dawkins, who has a gift for writing eloquent, quotable phrases, followed this with a series of other very readable works explaining for a popular audience how Darwinian natural selection provides the mechanism by which complex organisms evolve from simpler forms. In *The Extended Phenotype* he basically carried forward the idea of selfish genes.[35] In *The Blind Watchmaker* he responds specifically to the argument from design that had been compellingly presented by Archdeacon William Paley in *Natural Theology*, published fifty years before Darwin.[36]

In *River out of Eden* Dawkins makes evolution even more accessible to the popular audience.[37] *Climbing Mount Improbable* adds to this, showing how natural selection is able to produce systems that have apparently low *a priori* probability, answering creationist claims that biological organisms are so highly improbable that they cannot have arisen from pure natural processes.[38]

Dawkins switches gears a bit in *Unweaving the Rainbow*, where he shows that science has a beauty that matches art and poetry and should be looked at equally as a source of pleasure and inspiration.[39] *A Devils' Chaplain* is a series of essays on various subjects from pseudoscience to religion and creationism.[40] In *The Ancestor's Tale*, Dawkins and his assistant Yan Wong trace the evolutionary history of humanity backward in time and meet up with various other species who share the same ancestors.[41]

Finally we come to one of the publishing sensations of recent years, Dawkins's *The God Delusion*.[42] As of November 2007 it had sold 1.5 million copies and had been translated into thirty-one languages. It remained on the *New York Times* Hardcover Nonfiction Best Seller List for fifty-one weeks, until September 30, 2007. The paperback edition also became an instant and long-lasting best seller.

Dawkins argues that the "God hypothesis" is a valid scientific hypothesis that should be analyzed skeptically like any other. He goes over the arguments for God's existence and concludes they are "spectacularly weak." Referring to the argument from design, perhaps the most common reason given for belief in God, Dawkins says, "Far from pointing to a designer, the illusion of design in the living world is explained with far greater economy and with devastating elegance by Darwinian natural selection."[43]

Dawkins presents the case that religion is not the source of morality and is, in general, not such a good thing for the world. He claims that childhood indoctrination is the reason for much religious belief, rather than the free choice of a mature, rational evaluation of the alternatives.

As in previous books, Dawkins tries to show that "a proper understanding of the magnificence of the real world, while never becoming a religion, can fill the inspirational role that religion has historically—and inadequately—usurped."[44]

Finally, Dawkins urges that atheists not be shy or apologetic about their atheism but that they should stand tall to face the far horizon and speak out, proudly showing their independence of mind. In the past century in America we have seen women, then African Americans and other minorities, and then homosexuals stand up for themselves as humans with the same rights as others. Gradually these groups have become increasingly respected and accepted as equal members of society. It is time for atheists to gain the same status.

By far most of the books written to counter the rise of New Atheism have focused on *The God Delusion.* Let's see what a somewhat random selection of them have to say.

In *The Dawkins Delusion,* Alister and Janna Collicut McGrath assert, "Dawkins simply offers the atheist equivalent of slick hellfire preaching, substituting turbocharged rhetoric and highly speculative manipulation of fact for careful, evidence-based thinking."[45] Continuing,

> How . . . could such a gifted popularizer of the natural sciences, who once had such a passionate concern for the objective analysis of evidence, turn into such an aggressive antireligious propagandist with an apparent disregard for evidence that was not favorable to his case? Why were the natural sciences being so abused in an attempt to advance atheist fundamentalism?[46]

When I read this I made mental note to record, as I read through the rest of the McGraths' book, the many evidence-based examples of this abuse I expected the authors to give. But they simply say, "The book [*The God Delusion*] is often little more than an aggregation of convenient factoids suitably overstated to suggest that they constitute an argument. To rebut this highly selective appeal to evidence would be unspeakably tedious."[47]

So the McGraths could not bring themselves to overcome their tedium and give examples after all. As with many of the other books critical of New Atheism, the McGraths are guilty of the same scholarly laziness they attribute to the new atheists.

In *The Devil's Delusion,* David Berlinski does quote Dawkins and even takes the trouble (that must have taken some exertion) to give the page number:

> [Dawkins says] The God of the Old Testament is arguably the most unpleasant character in all of fiction: jealous and proud of it; a petty, unjust, unforgiving control-freak; a vindictive, blood thirsty ethnic cleanser; a misogynistic, homophobic, racist, infanticidal, genocidal, filicidal, pestilential, megalomaniacal, sadomasochistic, capriciously malevolent bully.[48]

Berlinski responds, "These are, to my way of thinking, striking points in God's *favor*, but opinions, I suppose, will vary."[49] It should be noted that Berlinski identifies himself as a "secular Jew."

John Haught, the theologian mentioned earlier, makes an astonishing statement that I will have more to say about in this book: "[Dawkins's] uncompromising literalism is nowhere more obvious than his astonishing insistence throughout *The God Delusion* that the notion of God should be treated as a scientific hypothesis."[50]

This pinpoints one of the key differences between atheists and theists as well as agnostics and many old atheists. The nonatheist groups seem to think that God is immune from being studied by the objective, rational methods of science. The new atheists firmly insist that the personal, Abrahamic God is a scientific hypothesis that can be tested by the standard methods of science. And, as we will see, he has failed the test.

Theists were not the only ones unhappy with *The God Delusion.* One of Dawkins's own colleagues in evolutionary biology, David Sloan Wilson, also an atheist, complains that Dawkins is wrong about religion:

> In *Darwin's Cathedral* I attempted to contribute to the relatively new field of evolutionary religious studies. When Dawkins' *The God Delusion* was published I naturally assumed that he was basing his critique of religion on the scientific study of religion from an evolutionary perspective. I regret to report otherwise. He has not done any original work on the subject and he has not fairly represented the work of his colleagues.[51]

Wilson does not review *The God Delusion* at all but spends his time arguing about evolutionary theories of religion, in particular the controversial question of group selection, which is not what Dawkins's book is about. Dawkins has no trouble shooting Wilson down:

> Why would Wilson "naturally assume" any such thing? Reasonable, perhaps, to assume that I would pay some attention to the evolution of reli-

gion, but why base a critique on an evolutionary perspective, any more than on Assyrian woodwind instruments or the burrowing behaviour of aardvarks? *The God Delusion* does, as it happens, have a chapter on the evolutionary origins of religion. But to say that this chapter is peripheral to my main critique would be an understatement. When I was asked to prepare an abridgment for the British audio recording, I had to decide which bits of the book were essential, and which bits could, however regretfully, be left out. My first cut, and the only chapter I deleted completely, was the chapter on evolutionary origins. Sad as I was to lose it (I was consoled by the fact that we also recorded an unabridged version for the American market) it seemed to me the least essential chapter to the central theme of the book.

BREAKING THE TABOO

Not all new atheists take as hard a line, and so are not as subject to as much God-lover venom as Harris, Dawkins, and Christopher Hitchens. We will get to Hitchens in a moment. But first let me mention the work of the eminent philosopher Daniel C. Dennett that has been considered part of New Atheism.

Dennett is the author of a dozen books. He is best known for two lengthy, scholarly tomes that present a strong case for the purely material world that I have identified as the atheist worldview. In *Consciousness Explained* (1991), Dennett attempted to show that the phenomenon we call consciousness can be understood as resulting from purely physical processes in the brain.[52] While a number of other scholars have made the same claim, Dennett's book is noted for arguing that the philosophical concept of *qualia*, which refers to the properties of sensory experiences—qualities or sensations such as pleasure or color—is incoherent and thus meaningless. This conclusion is, to say the least, highly controversial.

In *Darwin's Dangerous Idea* (1995), Dennett asserted that Darwinian natural selection not only is sufficient to explain the evolution of life but also is an organizing principle that can be applied in other domains such as mind, culture, morality, and society.[53] Also not exactly noncontroversial.

Dennett's contribution to the new atheist literature is called *Breaking the Spell: Religion as a Natural Phenomenon.* Unlike the other books we are summarizing, *Breaking the Spell* is not an argument for atheism and, although an outspoken atheist, Dennett insists he is not speaking to the choir. Rather he wants to convince believers and nonbelievers alike that, whether or not God exists, religion is such an important part of society that it should be studied scientifically, like any other natural phenomenon.

Many readers have misinterpreted Dennett's title as "breaking the spell of religion." However, he insists this is not what he meant. As he explains, "The spell that I say *must* be broken is the taboo against a forthright, scientific, no-holds-barred investigation of religion as one natural phenomenon among many."[54]

Furthermore, Dennett does not claim to have the answers to the questions that such a study would raise and all he does is make some tentative proposals in that direction.

Even his definition of religion is tentative: "I propose to define religions as *social systems whose participants avow beliefs in a supernatural agent or agents whose approval is to be sought.*"[55]

As many critics have pointed out, religion is already a subject of scientific study, so Dennett is really not proposing that much new. He summarizes the various proposals already found in a rapidly growing literature on the possible evolutionary origins of religion, including his own tentative scheme in which the unit of social evolution is not the gene but the "meme." *Meme* is a term introduced by Dawkins for any idea or thought that can replicate from one brain to another analogous to the biological gene. Dennett does not claim any of these theories, including his own, is established.

The biggest theist objection to Dennett, however, is again the one we saw before by John Haught—the simple act of applying science to religion. In his review in the *New York Times*, the literary editor of the *New Republic*, Leon Wieseltier, accuses Dennett of *scientism*:

> Scientism, the view that science can explain all human conditions and expressions, mental as well as physical, is a superstition, one of the dominant superstitions of our day; and it is not an insult to science to say so. For a sorry instance of present-day scientism, it would be hard to improve on Daniel C. Dennett's book. "Breaking the Spell" is a work of considerable historical interest, because it is a merry anthology of contemporary superstitions.[56]

The *Times* should not be asking a literary editor to review a book on science any more than they should ask a scientist like myself to review a book on Shakespeare. Wieseltier does not provide a quotation from Dennett, or any other new atheist for that matter, which asserts that science has all the answers. His is the typical misunderstanding of science found among academics who never took a course in science during all their years of education. Science provides many answers that are amply verified in the modern world and to refer to these as "contemporary superstitions" is an insult to scientists everywhere. Does Wieseltier go to a barber rather than a dentist when he has a toothache? Does he write his reviews on clay tablets instead of a word processor? Religion is a social phenomenon that is observable and thereby amenable to study by scientific means just as tribal customs in the Amazon or the culture of the Internet.

However, I do agree with Wieseltier when he says, "You cannot disprove a belief unless you disprove its content."[57] Dennett insists he is not interested in proving or disproving the existence of God and that the issue is not relevant to the scientific study of religion. His critics don't believe his claimed disinterest, and other new atheists agree that the existence or nonexistence of the Judeo-Christian-Islamic God is not only provable both logically and empirically but that his existence has been disproved beyond all reasonable doubt.

Still, it is hard to see how Dennett's book should have offended so many theists. I think his unpopularity among believers is more the result of his body of work including the books mentioned and various articles in newspapers and magazines. He and Richard Dawkins have been supportive of the movement among a group of atheists to use the term "brights" to refer to themselves in the same way that homosexuals refer to themselves as "gays." Dennett's op-ed piece in the *New York Times* pushing the bright idea drew a lot of attention.[58] The implication that offends most believers is that they are thereby "dims." Dennett denies this, noting that non-gays are not called "glums" but "straights" and believers could call themselves "supers" since they believe in the supernatural.[59] However, the term "bright" has not been adopted by the majority of new or old atheists. Christopher Hitchens has called it "conceited" and "cringe making."[60]

GOD: A FAILED HYPOTHESIS?

The next new atheist book to appear chronologically was my own effort, *God: The Failed Hypothesis—How Science Shows That God Does Not Exist*, which made the *New York Times* Best Seller List in March 2007.[61] There I argue that the Judeo-Christian-Islamic God can be proved not to exist beyond a reasonable doubt.

While I present some of the logical arguments that have appeared in recent literature, my main message is based on the fact that, although there are differences, the god of each of the three great monotheisms plays such an important role in the working of the universe and in the lives of humans that the effects of that deep involvement should be observable by the human senses and the instruments we have built to increase the power of these senses.

I disagreed sharply with those, including many nonbelieving scientists, who insist that science has nothing to say about God. Certainly science is not in a position to answer metaphysical questions about the nature of God, such as whether he exists outside of time. But it can examine the physical phenomena that should follow from the hypothesis of a God who is supposed to perform certain acts, from creating the universe and designing life, to answering prayers and revealing universal truths to humanity. These phenomena are testable just as those from any scientific theory. The failure of these tests implies the falsification of the hypothesis. The God hypothesis has failed those tests.

Now, what have the apologists said about my book? It should come as no surprise to hear that I did not convince Dinesh D'Souza. He refers to my cosmological discussion: "Physicist Victor Stenger says the universe may be 'uncaused' and may have 'emerged from nothing.'"[62] D'Souza scoffs, "Even David Hume, one of the most skeptical of all philosophers, regarded this position as ridiculous…Hume wrote in 1754, 'I have never asserted so absurd a proposition as that anything might rise without cause.'"[63]

Hume can be excused from not knowing quantum mechanics in 1754; but D'Souza cannot be excused in 2007, over a century since its discovery. According to conventional interpretations of quantum mechanics, nothing "causes" the atomic transitions that produce light or the nuclear decays that produce nuclear radiation. These happen spontaneously and only their probabilities can be calculated. In the course of this book we will see many more examples of the sheer ignorance or deliberate misrepresentation of science among Christian apologists such as D'Souza. We will also get into the more serious, knowledgeable objections when we discuss the various contentious issues in detail.

IS GOD GREAT?

The final book in the first round of new atheist literature that I will introduce here is *God Is Not Great: How Religion Poisons Everything* by Christopher Hitchens, which made first place on the *New York Times* Best Seller Hardcover Nonfiction List in early June 2007.

Hitchens is in many ways the most interesting of the new atheists, and is considered by many to be even more controversial than Richard Dawkins. Only theists hate Dawkins. Hitchens's detractors are more widely dispersed.

Hitchens was born in England and schooled in philosophy, politics, and economics at Oxford. He immigrated in 1981 to the United States, where he has been a prolific essayist and book reviewer after an earlier career as a foreign correspondent. He currently works for *Vanity Fair* magazine and his articles appear regularly in many other major journals.

Hitchens's political opinions are, to say the least, curious, as he has moved from a far-left Trotskyite to a supporter of many neoconservative positions. He has severely criticized Bill Clinton, Henry Kissinger, and Ronald Reagan. He has hit hard at religious leaders such as Mother Teresa, Jerry Falwell, and Tenzon Gyatso, the fourteenth Dalai Lama. His friendship with Salman Rushdie has led him to speak out forcefully against "Islamic fascism" and support the war in Iraq as part of the battle against this particular menace. As is the case with Sam Harris and all other new atheists, I will not discuss Hitchens's politics or any other opinions that have nothing to do directly with atheism.

"God is not Great" is an allusion to the Islamic exclamation "*Allahu Akbar*," or "God is Great." Hitchens lists four "irreducible objections" to religious faith:

- It wholly misrepresents the origins of humans and the cosmos.
- It manages to combine the maximum of servility with the maximum of solipsism.

- It is both the result and the cause of dangerous sexual repression.
- It is ultimately grounded on wish-thinking.[64]

In an epigraph Hitchens quotes John Stuart Mill on Mill's father's view of religion. It is worth repeating in full because it well represents the views of the new atheists and how they dramatically differ from the more conciliatory attitudes of the current mainstream of nonbelievers:

> His aversion to religion, in the sense usually attached to the term, was of the same kind with that of Lucretius: he regarded it with the feelings due not to a mere mental delusion, but to a great moral evil. He looked upon it as the greatest enemy of morality: first, by setting up factitious excellencies—belief in creeds, devotional feelings, and ceremonies, not connected with the good of human kind—and causing these to be accepted as substitutes for genuine virtue: but above all, by radically vitiating the standards of morals; making it consist in doing the will of a being, on whom it lavishes indeed all the phrases of adulation, but whom in sober truth it depicts as eminently hateful.[65]

God Is Not Great is available as an audio book read by the author, whose speaking voice is hardly distinguishable from that of the late, great actor Richard Burton.[66] Listening to chapter 4 you will hear an eloquent statement of the moral evils of religion, the unnecessary death and suffering of millions of people brought about by lunatic religious beliefs that are enforced by cowardly politicians throughout the world.

Theologian Haught admits: "The catalog of evils committed under the umbrella of theistic faiths is a long one, and the writings of Hitchens, Dawkins, and Harris could well serve as an examination of conscience by those of us who think of God as infinite goodness, self-serving love, the ground of our freedom, the author of life, and our ultimate destiny."[67]

Haught's response is not to seek a reason for the catalog of evils, but to ponder, "Not everybody thinks of God in such [infinitely good] terms, and it could be enlightening to find out why."[68] It would be more enlightening to find out why believers can still think of God in such loving terms given the catalog of evils that is presented in the three books, facts that in all cases cannot be disputed.

But to a theologian who works from the unquestioned starting assumption that an all-good, all-powerful, all-knowing God exists, the atheist has to be the one in error. He concludes that the atheist cannot find a solid justification for his ethical values in the absence of God. He refers to Harris as conjecturing that we can fall back on reason alone. Haught asks, "Why should we trust our reasoning abilities either?"[69]

Earlier we saw that Haught objected to treating God as a scientific hypothesis. His distrust in reason matches his distrust in science. And this is exactly the place where the battle line between theism and atheism is to be drawn. The theist thinks that he has some superior channel to reality, provided by God's revelations. The atheist argues that empirical science and reason are the most reliable tools we have to determine truths about the world. The reason we trust reason and science, and have no trust whatsoever in religious arguments, is that science and reason work in understanding the world and making it a better place for humanity while religious argument leads universally to dismal failure and untold human suffering.

NOTES

1. Victor J. Stenger, *Quantum Gods: Creation, Chaos, and the Search for Cosmic Consciousness* (Amherst, NY: Prometheus Books, 2008).

2. Richard Branham, private communication.

3. Bill Maher, *Religulous*, documentary film directed by Larry Charles, Lionsgate (2008).

4. The biblical story of Abraham is inconsistent with archaeological findings. See Israel Finkelstein and Neil Asher Silberman, *The Bible Unearthed: Archaeology's New Vision of Ancient Israel and the Origin of Its Sacred Texts* (New York: Free Press, 2001), pp. 33–36.

5. Michael Martin, "Atheism" in *The New Encyclopedia of Unbelief*, ed. Tom Flynn (Amherst, NY: Prometheus Books, 2007), p. 88.

6. Adherents.com, http://www.adherents.com/Religions_By_Adherents.html (accessed February 10, 2009).

7. David B. Barrett et al., *World Christian Encyclopedia: A Comparative Survey of Churches and Religions in the Modern World*, 2nd ed. (Oxford: Oxford University Press, 2001).

8. American Religious Identification Survey, Trinity College, http://www.americanreligionsurvey-aris.org/ (accessed May 5, 2009).

9. Fox News, November 1, 2008, http://elections.foxnews.com/2008/11/

01/dole-trails-democratic-rival-hagan-north-carolina-senate-race (accessed November 5, 2008).

10. The Baylor Religion Survey, Baylor Institute for Studies of Religion, September 2006. Selected findings at http://www.baylor.edu/content/services/document.php/33304.pdf (accessed May 12, 2008).

11. Dan Barker, *Godless: How an Evangelical Preacher Became One of America's Leading Atheists* (Berkeley, CA: Ulysses Press, 2008).

12. I will not give Web addresses for organizations or well-known individuals, since these frequently change and they can be easily found with an Internet search.

13. See Richard Dawkins, *The God Delusion* (Boston: Houghton Mifflin, 2008), pp. 375–79.

14. Stephanie Simon, "Atheists Reach Out—Just Don't Call It Proselytizing," *Wall Street Journal*, November 18, 2008.

15. Dinesh D'Souza, *What's So Great about Christianity?* (Washington, DC: Regnery, 2007), p. xv.

16. Becky Garrison, *The New Atheist Crusaders and Their Unholy Grail: The Misguided Quest to Destroy Your Faith* (Nashville: Thomas Nelson, 2007), p. 18.

17. John F. Haught, *God and the New Atheism: A Critical Response to Dawkins, Harris, and Hitchens* (Louisville, KY: Westminster John Knox Press, 2008), pp. 20–22.

18. Victor J. Stenger, *The Comprehensible Cosmos: Where Do the Laws of Physics Come From?* (Amherst, NY: Prometheus Books, 2006), pp. 304–12.

19. Sam Harris, *The End of Faith* (New York: Norton, 2004).

20. Ibid., p. 15.

21. Noam Chomsky, *9-11* (New York: Open Media/Seven Stories Press, 2001). Excerpts can be found at http://www.thirdworldtraveler.com/Chomsky/9-11_Chomsky.html (accessed November 12, 2008).

22. Fareed Zakaria, *The Future of Freedom: Illiberal Democracy at Home and Abroad* (New York: Norton, 2003), p. 138.

23. Harris, *The End of Faith*, p. 148.

24. Ibid., p. 18.

25. Ibid.

26. Beyond Belief: Science, Reason, Religion, and Survival, *Science Network*, http://thesciencenetwork.org/programs/beyond-belief-science-religion-reason-and-survival (accessed October 22, 2008).

27. Steven Weinberg, quoted by James Glanz, "Physicist Ponders God, Truth and 'A Final Theory,'" *New York Times*, January 25, 2000.

28. Steven Weinberg, *Freethought Today*, April 2000.

29. Garrison, *The New Atheist Crusaders and Their Unholy Grail*, p. 5.

30. Sam Harris, *Letter to a Christian Nation* (New York: Knopf, 2004).

31. Ibid., p. vii.

32. Sam Harris, "Response to Criticism," http://www.samharris.org/site/full_text/response-to-controversy2/ (accessed February 10, 2009).

33. Richard Dawkins, *The Selfish Gene* (New York: Oxford University Press, 1976).

34. There is no physical unit in a biological cell that is identified as a gene. Rather, it is a unit of specific information that is part of the DNA or RNA in a cell.

35. Richard Dawkins, *The Extended Phenotype: The Gene as the Unit of Selection* (Oxford; San Francisco: Freeman, 1982).

36. William Paley, *Natural Theology; Or, Evidences of the Existence and Attributes of the Deity*, 12th ed. (London: Printed for J. Faulder, 1809).

37. Richard Dawkins, *River out of Eden: A Darwinian View of Life*, Science masters series (New York: Basic Books, 1995).

38. Richard Dawkins, *Climbing Mount Improbable*, 1st American ed. (New York: Norton, 1996).

39. Richard Dawkins, *Unweaving the Rainbow: Science, Delusion, and the Appetite for Wonder* (Boston: Houghton Mifflin, 1998).

40. Richard Dawkins and Latha Menon, *A Devil's Chaplain: Selected Essays* (London: Weidenfeld & Nicolson, 2003).

41. Richard Dawkins, *The Ancestor's Tale: A Pilgrimage to the Dawn of Evolution* (Boston: Houghton Mifflin, 2004).

42. Dawkins, *The God Delusion.* The audio book, beautifully read by the author and his wife, actress Lalla Ward, is available in the United Kingdom from Random House, UK, and in North America by Tantor Media, Inc. (2006).

43. Ibid., p. 2.

44. Ibid., p. 3.

45. Alister McGrath and Joanna Collicutt McGrath, *The Dawkins Delusion: Atheist Fundamentalism and the Denial of the Divine* (Downers Grove, IL: InterVarsity Press, 2007), p. 11.

46. Ibid., p. 12.

47. Ibid., p. 13.

48. Dawkins, *The God Delusion.*

49. David Berlinski, *The Devil's Delusion: Atheism and Its Scientific Pretensions* (New York: Crown Forum, 2008), p. 44.

50. Haught, *God and the New Atheism*, p. 31.

51. David Sloan Wilson, *Darwin's Cathedral: Evolution, Religion, and the Nature of Society* (Chicago: University of Chicago Press, 2002).

52. Daniel C. Dennett, *Consciousness Explained* (London: Little, Brown, 1991).

53. Daniel C. Dennett, *Darwin's Dangerous Idea: Evolution and the Meanings of Life* (New York: Simon & Schuster, 1995).

54. Dennett, *Breaking the Spell*, p. 17.

55. Ibid., p. 9.

56. Leon Wieseltier, "The God Genome," *New York Times*, February 19, 2006.

57. Ibid.

58. Daniel C. Dennett, "The Bright Stuff, " *New York Times*, July 12, 2003.

59. Dennett, *Breaking the Spell*, p. 21.

60. Hitchens, *God Is Not Great*, p. 5.

61. Stenger, *God: The Failed Hypothesis*.

62. Victor J. Stenger, "Has Science Found God?" *Free Inquiry* 19, no. 1 (Winter 1998/1999): 56–58.

63. D'Souza, *What's So Great about Christianity?* p. 125; J. Y. T. Greid, ed., *The Letters of David Hume* (Oxford: Clarendon Press, 1932), p. 187.

64. Hitchens, *God Is Not Great*, p. 4.

65. John Stuart Mill on his father, in *Autobiography* (London: Longmans, Green, Reader, and Dyer, 1873), quoted in Hitchens, *God Is Not Great*, p. 15.

66. Hatchett Audio; unabridged ed., May 1, 2007.

67. Haught, *God and the New Atheism*, p. 72.

68. Ibid.

69. Ibid., pp. 73–74.

2. THE FOLLY OF FAITH

Faith is something that you believe that nobody in his right mind would believe.

—Archie Bunker

FAR FROM BENIGN

In America "people of faith" are treated with great deference. They are assumed to be persons of the highest moral standards—exemplars of goodness, kindness, and charity. But why should that be? How does faith qualify a person for such high esteem? After all, faith is belief in the absence of supportive evidence and even in light of contrary evidence. How can such a frame of mind be expected to result in any special insight?

In any human activity other than religion, someone ignoring evidence would be regarded as a fool. A homeowner would be unwise not to close up her house on the signs of an approaching storm. A shopkeeper would be imprudent to continue stocking an item that doesn't sell. What doctor, lawyer, detective, or scientist would go about his business without giving primary attention to evidence?

Yet, as Sam Harris says, "Criticizing a person's faith is currently taboo in every corner of our culture."[1] For example,

> When a Muslim suicide bomber obliterates himself along with a score of innocents on a Jerusalem street, the role that faith played in his action is invariably discounted. His motives must have been political, economic, or entirely personal. Without faith, desperate people would still do terrible things. Faith itself is always, and everywhere, exonerated.[2]

Christian apologist David Marshall quotes St. Paul: "Now faith is the substance of things hoped for, the *evidence* of things *unseen*."[3] Christopher Hitchens responds,

> If one must have faith in order to believe something, or believe *in* something, then the likelihood of that something having any truth or value is considerably diminished. The harder work of inquiry, proof, and demonstration is infinitely more rewarding, and has confronted us with findings far more "miraculous" and "transcendent" than any theology.[4]

One of the significant propositions of New Atheism is that faith should not be exonerated, should not be treated with respect, but rather disputed and, when damaging to individuals or society, condemned. In fact, we should call faith exactly what it is—absurd. The new atheists argue that faith is far from a benign force that can simply be tolerated by those who know better. Rather, it plays a significant role in much of the violent conflict in the world. Furthermore, faith results in the disregarding of important values such as freethinking and objective truth seeking that are needed to solve the problems in modern society. We have just lived through a disastrous eight-year period where decision after decision in the most important office in the world, the Oval Office of the White House, was made on the basis of an irrational mode of thought founded on faith and suspicious of any reasoned argument that contradicted that faith.

Harris asks, "Why did nineteen well-educated, middle class men trade their lives in this world for the privilege of killing thousands of our neighbors? Because they believed that they would go straight to paradise for doing so."[5]

Could anything have been more irrational, other than the responses of some Christian leaders that September 11 was God's punishment or the work of the devil?

DOING THE RIGHT THING

Even worse than September 11—and it is possible for something to be worse—belief in ancient myths joins with other negative forces in our society to keep most of the world from advancing scientifically, economically, and socially at a time when a rapid advancement in these areas and others is absolutely essential for the survival of humanity. We are now probably only about a generation or two away from the catastrophic problems that are anticipated from global warming, pollution, and overpopulation: flooded coastal areas, severe climatic changes, epidemics caused by overcrowding, and starvation for much of humanity. Such disasters are predicted to generate worldwide conflict on a scale that could exceed that of the great twentieth-century wars, with nuclear weapons in the hands of unstable nations and terrorist groups.

By virtue of its scientific, economic, and military predominance, America must necessarily lead the way out of these dangers. To do so will call for the best efforts of all elements of society, with science and technology leading the way by providing creative solutions and with business leaders and politicians implementing these solutions by uncharacteristically putting the commonweal ahead of their own selfish interests.

Unfortunately, while American science continues to predominate, it has begun slipping in many vital areas as the government has changed its funding priorities. The abysmal performance of American students in science and mathematics compared to other advanced nations does not bode well for the future.

As I write this, the world is in the middle of the worst economic crisis in years and many leaders are promising drastic change. An African American born in Hawaii with the unlikely name of Barack Hussein Obama has been dramatically elected president of the United States. Most of the world, myself included, have high hopes that they and we, working together with Obama, will be up to the task of solving the many problems we face. As Winston Churchill said, "You can always count on America to do the right thing... after they have exhausted all other possibilities."[6]

Whether or not we do the right thing remains to be seen. Let us take a look at the record of the policies of the past three decades that put us in the current situation. As we will find, the folly of faith played an important role. And while the promoters of so-called failed faith-based initiatives are now out of power, Obama says he will continue some of these initiatives. Furthermore, a sizable, organized Religious Right remains with huge financial resources, radio and TV stations, and publishing houses that will enable it to continue to influence events for years to come.

THE NEOCONS

In the decades since the Nixon administration, American business and government at the federal level have been strongly influenced, and in the George W. Bush administration eventually dominated, by an unholy alliance of *neoconservatives*, or *neocons*, and fundamentalist Christians whom author Damon Linker has dubbed *theoconservatives*, or *theocons*.[7]

Neoconservatism is a right-wing political movement that can be traced back to the writings of political philosopher Leo Strauss and the economic theories of Milton Friedman. It was adopted in the 1960s by anticommunist liberals who became disenchanted with the excesses of the left-wing counterculture of the time. Few, if any, liberals remain in the movement today. Neocons support most conservative values but differ from traditional conservatives in allowing big government and big deficits, and in promoting a more aggressive, go-it-alone foreign policy to protect and advance US interests. They have no qualms about seeing the development of an American empire that would not have to worry about world opinion. The neocon-theocon alliance helped keep an increasingly dogmatic Republican party in power in Washington for all but twelve of the last forty years. Notable neocons in the Bush administration included Paul Wolfowitz, Richard Pearl, Douglas Feith, Lewis Libby, John Bolton, Elliot Abrams, and Robert Kagan.

Historian Irving Kristol is regarded as the founder of neoconservatism.[8] His son is William Kristol, editor of the political magazine the *Weekly Standard*. In 1997, the younger Kristol and Robert Kagan founded the Project for the New American Century, a neoconservative think tank based in Washington, DC.

The influence of neoconservatives on the Bush administration became clear when their foreign policy ideas were adopted after the attacks of September 11, 2001, and renamed the *Bush Doctrine*. This doctrine led to the Iraq War and a general decline of US influence abroad as the need for international support for our actions was ignored and even scorned.

The collapse of the US economy in fall 2008 can also be attributed at least partially to neocon policies, although traditional conservatives, liberals, and both political parties can share the blame. Under the ideology of (faith in) unfettered free markets, many of the regulations that were installed during and after the Great Depression as part of Franklin Roosevelt's New Deal were dismantled. In the absence of regulation, greed for money and lust for power overwhelmed any sense of public good. And the people who suffered the most were those blue-collar Christians who had been bamboozled into voting for the "party of values," the Republicans.

One of the most unfortunate developments with neoconservatism was the cynical political move on its part to turn people away from a respect for education and rational thought. In this they fed off the anti-intellectualism that marked modern America in recent years.[9] Where once conservatives had a base among the highly educated, the neocons and their partner theocons decided to trade that base for the support of rural and working-class whites. In order to get them to vote against their own best economic interests, these groups were convinced that the intellectual elites, concentrated on both coasts, were not like them and did not understand their problems. Of course, the wealthy neocons didn't either, nor did they care. In the meantime, the resulting tax policies saw the richest 1 percent of American taxpayers receive a $1,000-per-week tax reduction, while the bottom 20 percent received $1.50 per week. Hopefully the specter of anti-intellectualism has been entombed with the election of Barack Obama.

THE THEOCONS

In the red states[10] of the Midwest and the South there was no longer any room for erudite coastal conservatives such as the late William F. Buckley, who spoke in complete sentences, his nose in the air, with a high-class accent. A Catholic in the Jesuit mold, he championed reasoned dialogue. There was certainly no room for scientific thinking in the administration of an inarticulate born-again Christian from Texas who claims he believed he was doing the work of God and based his decisions on faith rather than reason. Let us now see how these led America into disasters both home and abroad.

The theocons are influential figures from the Religious Right who share most neocon values but are more interested in social issues such as abortion and same-sex marriage than economics and foreign policy. They have the stated goal of pulling down the wall of separation between church and state, or simply claiming it does not exist, and converting America into a Christian theocracy. Their most important contribution to the unholy alliance is to rally Christians to go to the polls and support conservative candidates. (Three out of four white evangelicals voted for John McCain and Sarah Palin.)[11] Often this political activity crosses the legal and constitutional line that forbids churches and other tax-exempt nonprofit organizations from supporting individual candidates. This is yet another example of the special treatment given to religion in America. It is even allowed to break the law.[12]

A number of books have documented the theocon trend. Kevin Phillips, the highly respected author of the 2004 best seller *American Dynasty*,[13] reported in his 2006 book, *American Theocracy*, on the way the Bush administration allowed itself to be controlled by the coalition of neocons and theocons or, as he calls the latter, "religious zealots."[14] Phillips predicted the outcome we experienced with the economy in 2008: If left unchecked, the same forces will bring a preacher-ridden, debt-bloated, energy-crippled America to its knees."[15]

In her lively 2006 book, *Kingdom Coming*, reporter Michelle Goldberg refers to the movement to convert America to a Christian theocracy as *Christian nationalism*.[16] Journeying around America visiting classrooms, megachurches, and federal courts, she demonstrated how *dominionism*, the doctrine that Christians have the biblical-based right to rule nonbelievers, is threatening the foundations of democracy. Goldberg reports that members of the Christian nationalism movement occupied positions throughout the federal bureaucracy during the Bush years, "making crucial decisions about our national life according to their theology."[17] In those positions they allowed "knowledge derived from the Bible to trump knowledge derived from studying the world."[18]

In his 2006 book, *American Fascists: The Christian Right and the War on America*, journalist Chris Hedges told how the Christian Right pumps a warped version of a Christian America into tens of millions of homes through television, radio, and the curriculum of Christian schools.[19] As he says, "The movement's yearning for apocalyptic violence and its assault on dispassionate, intellectual inquiry are laying the foundation for a new, frightening America."[20] Hedges likens the theocon movement to that of the young fascists in Italy and Germany in the 1920s and '30s.

In *The Theocons*, Damon Linker shows how the Catholic Church joined hands with right-wing fundamentalist Protestants to work for three decades to inject their radical religious ideas into American politics—and how they succeeded. As an editor of the Catholic journal *First Things,* Linker was a witness to these developments firsthand.

Linker tells the story of how the neocons and theocons began to forge an alliance as early as the 1970s, motivated largely by the massive cultural changes of the times. The key figures on the theocon side were clergyman Richard John Neuhaus, who began as a left-wing political radical and gradually converted to the neocon view, and Catholic philosopher Michael Novak, who followed a similar path.

Originally an ordained Lutheran minister from Canada, Neuhaus saw the Catholic Church as a greater source of discipline and authority and so converted to Catholicism in 1990.[21] He was ordained a priest the following year. In 1990 Neuhaus founded *First Things,* which promoted much of the neocon-theocon agenda. He became a close, unofficial adviser to President George W. Bush.

Novak has written numerous books on capitalism and religion and was the winner of the 1994 Templeton Prize for Progress in Religion. He served as US chief ambassador to the United Nations' Commission on Human Rights in 1981 and has performed other diplomatic functions.

Novak's review of the new atheist literature appeared in the *National Review* in 2007.[22] Other than extolling the virtues of religion and voicing the usual complaint that Dawkins, Harris, and Dennett do not understand theology, Novak offers no meaningful counters to the new atheist arguments. He has followed up this ineffectual work in 2008 with a book, *No One Sees God: The Dark Night of Atheists and Believers.*[23] There he dismisses Harris, Dennett, Dawkins, and Hitchens as "difficult to engage" on religion since "all of them think that religion is so great a menace that they do not show much disposition of dialogue."[24] This enables him to make sure the dialogue is on his terms.

Under the spell of the theocons, George W. Bush relied for eight years on faith rather than reason to make decisions, such as invading Iraq, that affect every person on Earth. He used dogma rather than data to make policy. His administration ignored or rewrote scientific evidence to suit the demands of the Christian Right, who had convinced Bush he was doing God's work. I will mention some examples. More can be found in *The*

Republican War on Science by journalist Chris Mooney,[25] as well as the books I have already referenced.

The Bush administration was noted for its "faith-based" initiatives. Billions of dollars annually in federal funds were awarded directly, unconstitutionally, and without congressional approval to churches and other religious organizations.[26] Those organizations were allowed to break the law (and Constitution) and hire only those of the same faith. Independent studies have shown that faith-based organizations do no better in providing needed social services than secular ones and often use the money unlawfully for religious purposes such as proselytizing. President Obama has said he would continue the program while curbing its abuses.[27]

Perhaps more than any other, the Bush administration ignored the advice of those who disagreed with its ideology. In May 2004, the Food and Drug Administration refused to approve over-the-counter sales of the "morning after" pill Plan B, ignoring a 23–4 recommendation from its scientific advisers.[28] Main opposition came from W. David Hager, an obstetrician and gynecologist who blends religion and medicine, endorsing the alleged healing power of prayer and prescribing Bible readings for the treatment of premenstrual syndrome.[29]

Also, some Christian preachers claim that abortions cause breast cancer. In 2002 the National Cancer Institute was forced to remove a fact sheet from its Web site that said there was no connection.[30]

These are just two specific examples. In 2004 the Union of Concerned Scientists issued a report charging the Bush administration with "manipulation of the process through which science enters into its decisions."[31] This report was signed by 12,000 scientists including Nobel laureates and other eminent scholars. It included detailed documentation of specific instances of abuse. I will just list the general findings:

- There is a well-established pattern of suppression and distortion of scientific findings by high-ranking Bush administration political appointees across numerous federal agencies. These actions have consequences for human health, public safety, and community well-being.
- There is strong documentation of a wide-ranging effort to manipulate the government's scientific advisory system to prevent the appearance of advice that might run counter to the administration's political agenda.

- There is evidence that the administration often imposes restrictions on what government scientists can say or write about "sensitive" topics.

TOWARD THE APOCALYPSE

If this is not sufficiently alarming, an influential element of the Religious Right have managed to convince a large segment of the American people that we are heading toward the end times with the Second Coming of Christ prophesied in the book of Revelation, the final book of the New Testament. As Phillips describes them: "The rapture, end-times, and Armageddon hucksters in the United States rank with any Shiite ayatollah, and the last two presidential elections mark the transformation of the GOP into the first religious party in U.S. history."[32] Millions of Americans take seriously the end-times scenario proposed by Hal Lindsey in his 1970 best seller, *The Late Great Planet Earth*, which has sold over 30 million copies.[33] More recently, Tim LaHaye and Jerry B. Jenkins with their Left Behind series of novels, which first appeared in 1995, had by 2005 sold no less than 65 million copies.[34] The series begins with three prequels to the Rapture, an event that does not appear in Revelation and is apparently based on a poetic reference by Paul in Epistle 1 to the Thessalonians:

> For the Lord himself shall descend from heaven with a shout, with the voice of the archangel, and with the trump of God: and the dead in Christ shall rise first: then we which are alive *and* remain shall be caught up together with them in the clouds, to meet the Lord in the air: and so shall we ever be with the Lord. (1 Thess. 4:16–17, KJV)

The Rapture described in *Left Behind* is not part of traditional teaching in Christianity. In the novel, airline passengers suddenly disappear from their seats and cars veer driverless down the streets as good Christians are whisked off to heaven so they need not suffer the upheaval to follow. I am sure you have all seen the bumper stickers "In Case of Rapture This Car Will Have No Driver" and the counter, "In Case of Rapture, Can I Have Your Car?" It's all fiction, of course, but then so is most of the Bible. Two-thousand-year-old mythical tales are no more reliable than those written today. Indeed, less.

The book of Revelation was written on the tiny Greek island of Patmos by an exile usually identified as John the Apostle, who was almost certainly not the disciple John or the author of the Gospel of John. (They are also unlikely to be the same person.)

In a series of imaginative visions termed the Apocalypse, John of Patmos details the Second Coming of Jesus in which Christ leads the fight against Satan, also known as the Antichrist and the Beast, culminating in the great battle at Armageddon (Megiddo in Israel, the site of many historical battles), where Satan and his forces are annihilated.

After Armageddon Jesus and the saints rule his kingdom on Earth for a thousand years. Satan somehow reappears and starts more mischief, but his armies are defeated by a judgment of fire from heaven. That final judgment is applied to all who have ever lived, with the evil cast into the "Lake of Fire" that is the new hell. Then begins the renewal of the entire Creation—a new heaven and a new Earth.

> See, the tabernacle of God is among humans!
> He will make his home with them,
> and they will be his people.
> God himself will be with them,
> and he will be their God.
> He will wipe every tear from their eyes.
> There won't be death anymore.
> There won't be any grief, crying, or pain,
> because the first things have disappeared. (Rev. 21:1–4, International Standard Version)

Although apocalyptic thinking is not common in Catholicism and mainline Protestantism, biblical historian Bart Ehrman argues that it is the most significant idea in the development of both Judaism and Christianity.[35] Ehrman claims the evidence indicates that Jesus (if he existed at all) was one of many apocalyptic preachers in the ancient world who promised that suffering and death would end when some agent of God would come down from heaven—Jesus called him the "Son of Man"—and institute the Kingdom of God on Earth.[36] Jesus is recorded as saying "... there be some standing here, which shall not taste of death, till they see the Son of Man coming in his kingdom" (Matt. 16:28, KJV). He makes similar predictions in four other places in the Gospels (Mark 9:1; Mark 13:30;

Matt. 24:34; Luke 9:27). When the coming did not happen within the lifetimes of his disciples, as Jesus prophesied, Christianity changed its emphasis to the Resurrection and promise of eternal life.

Still, throughout history Christians have expected the imminent coming of Christ (assumed to be the Son of Man). Every failed prophecy of a specific date was followed by yet another prophecy. Many predictions of the end occurred approaching special years such as 1000 CE, 1500 CE, and, we can all recall, 2000 CE. The eighteenth and nineteenth centuries saw a rise in *millennialism* in which a final battle between good and evil takes place, ushering in the millennium of Jesus' rule. The Seventh-Day Adventist Church was founded by William Miller, who predicted the Second Advent of Jesus in or around 1844. The church has survived the failure of this prophecy by continually recalculating it.

Other Christian sects that preach an approaching apocalypse include the Pentecostals, Jehovah's Witnesses, Mormons, and Rastafarians. Modern evangelicals are taught a form of apocalypticism called *dispensationalism* that differs from the traditional millennialism in including the Rapture. Outside Christianity, Islam, Zoroastrianism, Buddhism, Hinduism, the Baháh'í Faith, and the Hopi, Lakota, Mayan, and other Native American religions also speak of an end to the world as we know it to be followed by a purified world of harmony and spiritual peace.

Obviously, a transition to a better world, one without death or suffering, has great appeal. And the fact that it can be found in virtually all ancient and modern religious traditions gives the notion the veneer of sacred authority.

But there is nothing under that veneer to provide any basis for such hopes. Once again it is faith without evidence. Indeed, the complete failure of any of the dated prophecies of the Second Coming, notably those prophecies attributed to Jesus himself, provides solid evidence that an ultimate heaven on Earth is not going to happen anytime in the future.

Yet many Americans actually believe the Apocalypse is going to happen, and happen soon. A 2006 Pew survey found that 79 percent of all Christians believe in the Second Coming, with 20 percent believing they will see it in their lifetimes. The latter includes 33 percent of all white evangelicals, 34 percent of all black Protestants of all sects, 7 percent of white mainline Protestants, 12 percent of all Catholics, and 8 percent of all white non-Hispanic Catholics.[37]

Given that apocalyptic thinking is so widespread in the United States, there may someday be a president who thinks that the world is coming to an end in his or her administration. Such a president would see no reason to worry about global warming or other environmental problems, or even nuclear holocaust, since these would be part of the events leading to the Second Coming. She might even stir up violence in the Middle East to get things moving along toward the Final Judgment.

If you think this is far-fetched, just consider the 2008 presidential election. In the last weeks of the campaign, some Christians were so desperate to change the trend for Obama that they tried every dirty trick in the book to convince people that he was too dangerous to put in office. (Good Christians, all.) On one conservative Web site called "News By Us," Rev. Michael Bresciani wrote: "I can't fight off the inclination to believe that if their [*sic*] was a person alive today that looks more like the false prophet spoken of in the Bible; it is Barack Obama."[38]

While the Bush administration is now out of power and Barack Obama has a mandate to overturn Bush policies, the Christian Right still remains a major force in America with its huge financial resources and thousands of media outlets. While the new president has stated his strong support for the separation of church and state, the Republican candidate for vice president in 2008, then governor Sarah Palin of Alaska, is a member of the Assemblies of God, a Pentecostal Christian church. In their statement of beliefs, numbers 14–16 read:

> WE BELIEVE . . . in The Millennial Reign of Christ when Jesus returns with His saints at His second coming and begins His benevolent rule over earth for 1,000 years. This millennial reign will bring the salvation of national Israel and the establishment of universal peace.
>
> WE BELIEVE . . . A Final Judgment Will Take Place for those who have rejected Christ. They will be judged for their sin and consigned to eternal punishment in a punishing lake of fire.
>
> WE BELIEVE . . . and look forward to the perfect New Heavens and a New Earth that Christ is preparing for all people, of all time, who have accepted Him. We will live and dwell with Him there forever following His millennial reign on Earth. "And so shall we forever be with the Lord!"[39]

Palin could very well be the 2012 Republican presidential nominee.

All these Christians looking forward to the End of Days should pick up their Bibles and read the prophecies of John a little more carefully. The first thing that will happen is that 144,000 male Jews will be raised from the dead and have YHWH branded on their foreheads:

> Saying, Hurt not the earth, neither the sea, nor the trees, till we have sealed the servants of our God in their foreheads. And I heard the number of them which were sealed: and there were sealed an hundred and forty and four thousand of all the tribes of the children of Israel. (Rev. 7:3–4)

Then come the locusts:

> And there came out of the smoke locusts upon the earth: and unto them was given power, as the scorpions of the earth have power. And it was commanded them that they should not hurt the grass of the earth, neither any green thing, neither any tree; but only those men which have not the seal of God in their foreheads. (Rev. 9:3–4)

My guess is that all women, who never get a break in the Bible, are wiped out as well, so only 144,000 Jewish males remain. Then Jesus appears in the form of a lamb:

> And I looked, and, lo, a Lamb stood on the mount Sion, and with him an hundred forty [and] four thousand, having his Father's name written in their foreheads. (Rev. 14:1)

And these are not a representative sample of Jewish men:

> . . . the hundred and forty and four thousand, which were redeemed from the earth. These are they which were not defiled with women; for they are virgins. (Rev. 14:3–4)

Well, this does not much resemble the *Left Behind* scenario. According to the Bible not a single living human is saved, only 144,000 male Jewish virgins resurrected from the dead.

FAITH AND EVIDENCE

Several Christians have told me personally that Christianity does not teach the wisdom of "blind faith" but always associates faith with evidence. But then when I ask them to give me an example of such evidence they say things like, "Well, there was the empty tomb."

There is not a single piece of independent historical evidence for the existence of Jesus or the veracity of the events described in the New Testament.[40] Even the much-touted statement by the Jewish historian Flavius Josephus is now accepted by almost all scholars as a forgery. The paragraph in *Antiquities* that mentions Christ, his "wonderful works," death on the cross, and appearance three days later does not appear in earliest copies of that work and not until the fourth century.[41]

A number of scholars have made the case for the nonhistoricity of Jesus, and their conclusions are convincing.[42] As I keep saying, absence of evidence is evidence of absence when the evidence should be there and is not. In the case of Jesus, there were several historians living in or near Judea at the time who reported on all kinds of events there, but they never mention Christ.

One example is Philo-Judaeus, also known as Philo of Alexandria. In *The Christ,* John E. Remsberg writes:

> Philo was born before the beginning of the Christian era, and lived long after the reputed death of Christ. He wrote an account of the Jews covering the entire time that Christ is said to have existed on earth. He was living in or near Jerusalem when Christ's miraculous birth and the Herodian massacre occurred. He was there when Christ made his triumphal entry into Jerusalem. He was there when the crucifixion with its attendant earthquake, supernatural darkness and resurrection of the dead took place—when Christ himself rose from the dead and in the presence of many witnesses ascended into heaven. These marvelous events which must have filled the world with amazement, had they really occurred, were unknown to him.[43]

David Marshall gives the oft-heard argument that Christ gave sufficient evidence to his disciples that they were willing to die for him.[44] But that's a story that could easily be pure fiction.

Early Christianity was not the only time in history when people have died for a cause. Consider the Japanese and their code of Bushido. I do not doubt that many followers of the myth of Jesus gave up their lives for it—even if he didn't exist. But that does not prove he is God. Many German soldiers sacrificed their lives for Hitler in World War II and that did not make him God. The kamikaze pilots in World War II are perhaps analogous to Christ's disciples. To them the emperor was God, but we know, and he admitted, that he was a human just like the rest of us.

Marshall tries to argue that evidence from the Bible is no less credible than evidence in science:

> Almost everything we know—not just about first-century Palestine, but about dwarf stars, neutrinos, state capitals, vitamins, and sports scores—we believe because we find the person telling us the information is credible.[45]

Yes, but the stories of the Bible are *incredible.* Isn't it incredible that someone rose from the dead? To believe that requires far more evidence than a ball score in the newspaper. And, as someone who labored for thirty years to learn the properties of neutrinos, I can tell you that the evidence for their existence far exceeds any evidence that someone rose from the dead.

Marshall tries to elaborate the meaning of faith:

> Faith involves a continuum of four kinds of trust. First, we trust our minds. There's no way to prove our minds work—this is often forgotten by people who uncritically praise the scientific method. Even to do math or logic, which are more basic than science, we have to take our brains more or less for granted. How could we prove them? Any proof would depend on what it assumes: the validity of that endless electrical storm in the brain.[46]

We trust scientific method, logic, and mathematics because they work. They give us answers that we can independently test against objective observations. They give us electric lights, computers, and cell phones.

Science flies us to the moon. Religion flies us into buildings.

Marshall keeps to the same line of reasoning: "The second level of faith is trust in our senses. . . . Again there's no way to prove your eyes, ears, nose, mouth, and skin are giving you the real scoop about the outside world."[47]

True, we can't prove our senses are giving us the "real scoop." But we have plenty of personal experience that our senses do a good job of alerting us to oncoming cars, warning us when something on the stove has caught fire, and telling us that the baby needs to be fed.

Marshall turns to testimonial evidence: "Third, to learn anything we accept 'testimonial evidence' from parents, teachers, books, street signs, Wikipedia, and 'familiar' voices transmitted as electronic pulses over miles of wire and electromagnetic signals, then decoded into waves in the air. Almost everything we know comes from other people one way or another. This is true in science."

Yes, but we don't just take anyone's word for it. We test against independent observations. If I went up to a colleague and told him I solved some major physics problem, do you think he would simply accept that without insisting I prove it to him?

Of course we don't have time to independently test everything we hear, so we take the word of credible people. But that's because these people have already demonstrated their credibility by proving to be reliable in the past. That's why scientists and scholars of all kinds work so hard to maintain a good reputation. No one pays any attention anymore to Stanley Pons and Martin Fleischmann, the chemists who announced to the world in 1989 that they had discovered cold fusion.

It also depends on what is the message. If an airline pilot flying over Yellowstone National Park reports seeing a forest fire, we have no reason to doubt her. But if she reports seeing a flying saucer whose pilot waved a green tentacle at her, I would demand more evidence.

Besides, much testimonial evidence is highly unreliable, as demonstrated by the hundreds of death row inmates who were convicted by eyewitness testimony and later exonerated by DNA evidence in recent decades. Physical evidence is what matters the most.

Marshall goes on to accuse scientists of hubris: "The problem is that hubris about the 'scientific method' often masks an almost childish naiveté about what constitutes a good argument in nonscientific fields."[48] This is not naiveté at all but an objective evaluation of the relative success of science and nonscience. Marshall does not give one example of a nonscientific argument that has the same power as a typical scientific argument.

Finally we get to God: "In fact, scientific evidence is based on faith—exactly the same sort of faith as informed Christians have in God."[49] Scientific evidence must pass tests. As I detailed in my books *Has Science Found God?* and *God: The Failed Hypothesis*, evidence for a God such as the Abrahamic God fails all tests.[50]

Marshall then gets to the fourth level of faith: "The fourth level of faith is religious.... Faith must 'precede' reason, Augustine said, because there are some truths that we cannot yet grasp by reason."[51] Marshall does not name one truth obtained by faith that has been verified by independent, objective means. And that is not an impossible requirement. We can easily imagine some truth about the world obtained by spiritual revelation telling us something about the world that can be observed. For example, suppose such a truth is that only one religion is correct and the rest false. Then the prayers of the true religion should be answered while other prayers are not.

Next we get to the hidden God: "God could choose to remain hidden."[52] Sure, but such a God would never be known to anyone, as Marshall admits when he says: "But Christianity says God hasn't remained hidden. He reveals himself in creation, our hearts, and in history."[53]

Then why are there any non-Christians?

I discussed the hiddenness of God question in some detail in *God: The Failed Hypothesis*.[54] Philosophers John Schellenberg and Theodore Drange have independently shown how a moral God who deliberately hides himself from anyone open to evidence for his existence cannot logically exist.[55] The God of Evangelical Christians and several other sects, which is based on the teachings of John Calvin, permits only his special favorites to join him in heaven and damns everyone else no matter how saintly. This is a possible God, just not a moral God. Indeed, an evil God is fully compatible with the data.

Theologian John Haught has also objected to the new atheist treatment of faith as irrational. He tries to argue that religious faith is no different from what is assumed in other areas of discourse. He refers to the statement made in the 1960s by the eminent biochemist and atheist Jacques Monod:

> Monod claimed that the "ethic of knowledge" must be the foundation of all moral and ethical claims... it is unethical to accept any ideas that fail to adhere to the "postulate of objectivity." In other words, it is morally wrong to accept any claims that cannot be verified in principle by "objec-

> tive" scientific knowing. But, then, what about the precept itself? Can anyone prove objectively that the postulate of objectivity is true?[56]

Once again we see a theologian demonstrating his misunderstanding of science. The validity of the postulate of objectivity is not to be proven by some philosophical, deductive argument. Its validity is proved beyond a reasonable doubt by the empirical evidence of its methodological success.

Haught continues to show his bias against science:

> At some foundational level all knowing is rooted in a declaration of trust, in a "will to believe." For example, we have to trust that the universe makes some kind of sense before we begin the search for intelligibility. Unacknowledged declarations of faith underlie every claim the atheist makes as well, including the repudiation of faith.[57]

This is not the way it works at all. What Haught calls "trust" is what scientists call a "working hypothesis." If that hypothesis leads to a theory that gives the wrong answers, then it is discarded. By contrast, the blind trust of religious faith continually leads to the wrong answers but is never discarded.

And once again we hear a scientifically challenged theologian say that science is based on faith:

> There is no way, without circular thinking, to set up a scientific experiment to demonstrate that every true proposition must be based on empirical evidence rather than faith.... The claim that truth can be attained only by reason and science functioning independently of any faith is itself a faith claim.[58]

On the contrary, every successful scientific experiment that results in a practical application demonstrates the utility of basing our theories on empirical evidence. As explained above, whether or not it is "true" in some metaphysical sense is irrelevant, as long as it works.

Haught makes an unwarranted claim about faith: "Theology thinks of faith as a state of self-surrender in which one's whole being, and not just the intellect, is experienced as being carried away into a dimension of reality that is much deeper and more real than anything that could be grasped by science and reason."[59] How does he know that this is not simply a delusion? It is an experience easily, and more reliably, produced by certain drugs. And why can't a "deeper dimension of reality" be grasped by science and reason?

And again, science is belittled: "Scientific method by definition has nothing to say about God, meaning, values, or purpose."[60] By whose definition? I will have a lot more to say about science and God in this book. But for now let me comment that science and reason can be applied to anything and everything that involves some sort of observation. This includes the "inner" observations we make in our minds.

We then move on to whether we can explain faith in Darwinian fashion, as suggested by Dawkins[61] and others:

> If Darwinian theory were exclusively explanatory of religious faith, there would be little reason to complain about it. Religion in that case would be just one more instance of the clumsy creativity of nature, no more objectionable than vestigial organs. But for Dawkins's religious faith is in every respect an ethically despicable development, so the blame must not fall on blind and morally innocent Darwinian mechanisms. For Dawkins, evolution itself is not evil but merely indifferent. The evil in religion must then be extraneous to the life process, and therefore out of the scope of biology to account for it.[62]

But evolution does produce evil, such as all the gratuitous suffering in nature. So why shouldn't it be capable of producing the evil of religion?

Haught admits that religions are capable of evil. He calls it "idolatry": "The antidote to idolatry ... is not atheism but faith."[63] Faith is the source of idolatry in the first place.

NOTES

1. Sam Harris, *The End of Faith: Religion, Terror, and the Future of Reason* (New York: Norton, 2004), p. 13.
2. Ibid.
3. Hebrews 11:1.
4. Christopher Hitchens, *God Is Not Great: How Religion Poisons Everything* (New York: Twelve Books, 2007), p. 71.
5. Harris, *The End of Faith*, p. 29.

6. There are several versions of this quotation. Source unknown.

7. Damon Linker, *The Theocons: Secular America under Siege* (New York: Doubleday, 2006).

8. Irving Kristol, *Two Cheers for Capitalism* (New York: Basic Books, 1978); *Neoconservatism: The Autobiography of an Idea* (New York: Free Press, 1995).

9. Susan Jacoby, *The Age of American Unreason* (New York: Pantheon Books, 2008).

10. US media color Republican states red and Democratic states blue. This is opposite to the European convention where red is the color associated with the left wing. This was probably done to avoid any suggestion that either party was communist.

11. Pew Forum on Religion and Public Life, Exit Poll 2008 Presidential Election, http://pewresearch.org/pubs/1022/exit-poll-analysis-religion (accessed May 20, 2009).

12. Russell Goldman, "Pastors Challenge Law, Endorse Candidates From Pulpit," http://abcnews.go.com/Politics/Vote2008/Story?id=5198068&page=1 (accessed May 20).

13. Kevin Phillips, *American Dynasty: Aristocracy, Fortune, and the Politics of Deceit in the House of Bush* (New York: Penguin, 2004).

14. Kevin Phillips, *American Theocracy: The Peril and Politics of Radical Religion, Oil, and Borrowed Money in the 21st Century* (New York: Viking, 2006).

15. Ibid., book jacket.

16. Michelle Goldberg, *Kingdom Coming: The Rise of Christian Nationalism* (New York: Norton, 2006).

17. Ibid., p. 16.

18. Ibid., p. 127.

19. Chris Hedges, *American Fascists: The Christian Right and the War on America* (New York: Free Press, 2007).

20. Ibid., book jacket.

21. Linker, *The Theocons*, p. 82.

22. Michael Novak, "Lonely Atheists of the Global Village," *National Review*, March 19, 2007.

23. Michael Novak, *No One Sees God: The Dark Night of Atheists and Believers* (New York: Doubleday, 2008).

24. Ibid., p. 31.

25. Chris Mooney, *The Republican War on Science* (New York: Basic Books, 2005).

26. Goldberg, *Kingdom Coming*, p. 108.

27. Rob Boston, "Faith-Based Flare Up," *Church & State* 61, no. 8 (2008): 4–6.

28. Mooney, *The Republican War on Science*, p. 22.

29. Ibid., p. 216.

30. Ibid., p. 207.

31. Union of Concerned Scientists, "Investigation of the Bush Administration's Abuse of Science," http://ucsusa.wsm.ga3.org/scientific_integrity/interference/reports-scientific-integrity-in-policy-making.html (accessed October 11, 2008).

32. Phillips, *American Theocracy*, p. vii.

33. Hal Lindsey and Carole C. Carlson, *The Late Great Planet Earth* (Grand Rapids: Zondervan, 1970).

34. T. F. LaHaye and J. B. Jenkins, *Left Behind: A Novel of the Earth's Last Days* (Wheaton, IL: Tyndale House Publishers, 1995).

35. Bart D. Ehrman, *God's Problem: How the Bible Fails to Answer Our Most Important Question—Why We Suffer* (New York: HarperOne, 2008), p. 201.

36. Ibid., pp. 201–25.

37. Pew Research Center, "Many Americans Uneasy with Mix of Religion and Politics," http://pewforum.org/publications/surveys/religion-politics-06.pdf (accessed November 7, 2008).

38. Michael Bresciani, "Just Another Election or a Fight for the Soul of the Nation," *News by Us*, October 12, 2008, http://newsbyus.com/index.php/article/1770 (accessed October 18, 2008).

39. General Council of the Assemblies of God, "Our 16 Fundamental Truths Condensed," http://www.ag.org/top/Beliefs/Statement_of_Fundamental_Truths/sft_short.cfm (accessed October 12, 2008).

40. Dan Barker, *Godless: How an Evangelical Preacher Became One of America's Leading Atheists* (Berkeley, CA: Ulysses Press, 2008), pp. 261–76.

41. Ibid., pp. 254–59.

42. Earl Doherty, *The Jesus Puzzle: Did Christianity Begin with a Mythical Christ?* (Ottawa: Canadian Humanist Publications, 1999); George Albert Wells, *The Jesus Myth* (Chicago: Open Court, 1999); Robert M. Price, *Deconstructing Jesus* (Amherst, NY: Prometheus Books, 2000); Frank R. Zindler, *The Jesus the Jews Never Knew: Sepher Toldoth Yeshu and the Quest of the Historical Jesus in Jewish Sources* (Cranford, NJ: American Atheist Press, 2003).

43. John E. Remsburg, *The Christ: A Critical Review and Analysis of the Evidences of His Existence* (Amherst, NY: Prometheus Books, 1994).

44. David Marshall. *The Truth behind the New Atheism: Responding to the Emerging Challenges to God and Christianity* (Eugene, OR: Harvest House Publishers, 2007), p. 17.

45. Ibid., p. 18.

46. Ibid., p. 27.

47. Ibid., pp. 27–28.

48. Ibid., p. 28.

49. Ibid., p. 29.

50. Victor Stenger, *Has Science Found God?* (Amherst, NY: Prometheus Books, 2003); *God: The Failed Hypothesis—How Science Shows That God Does Not Exist* (Amherst, NY: Prometheus Books, 2008).

51. Marshall, *The Truth behind the New Atheism*, p. 30.

52. Ibid.

53. Ibid., p. 31.

54. Stenger, *God: The Failed Hypothesis*, pp. 237–41.

55. J. L. Schellenberg, *Divine Hiddenness and Human Reason* (Ithaca, NY: Cornell University Press, 1993); Theodore M. Drange, *Nonbelief & Evil: Two Arguments for the Nonexistence of God* (Amherst, NY: Prometheus Books, 1998).

56. John F. Haught, *God and the New Atheism: A Critical Response to Dawkins, Harris, and Hitchens* (Louisville, KY: Westminster John Knox Press, 2008), p. 5.

57. Ibid., p. 6.

58. Ibid., p. 11.

59. Ibid., p. 13.

60. Ibid., p. 18.

61. Richard Dawkins, *The God Delusion* (Boston: Houghton Mifflin, 2008), pp. 163–66.

62. Haught, *God and the New Atheism*, p. 59.

63. Ibid., p. 76.

3. THE SWORD OF SCIENCE

> **Every time that we say that God is the author of some phenomenon, that signifies that we are ignorant of how such a phenomenon was caused by the forces of nature.**
> —Percy Bysshe Shelley (d. 1822)

CAN SCIENCE STUDY THE SUPERNATURAL?

In the battle against superstition, no one wields a mightier sword than the scientist. No wonder, then, that much of the criticism of New Atheism is addressed to its reliance on scientific evidence and rational argument.

Christian apologist David Marshall asserts:

> The New Atheism reveals its simplistic grasp on reality in many ways. First, the most cocky atheists often fail to recognize the limits of science. Second, their theories leave too many facts out. Third, they refuse to ask certain obviously important questions. Fourth, to obscure the failure of their theory, some are driven to play a game of "let's pretend."[1]

Marshall does not quote a single cocky, or even noncocky atheist who says that science has no limits. In all our books and essays we talk about our appreciation of art, music, poetry, loving relationships, and all the many wonderful nonscientific gifts that nature and life give us.[2]

At the same time, we do not admit that scientific thinking makes no contribution to our appreciation of these endeavors. Understanding the physics of music helps in its appreciation, performance, and the manufacture of instruments. Recordings make it possible to enjoy music under many circumstances, such as while riding an exercise bike. Science helps detect art fraud and provides new visual art forms with the aid of computers. Soon every poem and novel ever written will be available for downloading from the Internet, which has already become invaluable to writers and scholars as an easily retrievable information source.

I haven't the faintest idea what Marshall is talking about in leaving facts out, refusing to ask important questions, and playing "let's pretend." He gives no examples.

Intelligent design spokesman David Berlinski tells us that scientists are "widely considered self-righteous, vain, politically immature, and arrogant."[3] He singles out my good friend and colleague Taner Edis and me, saying we "exhibit the salient characteristic of physicists endeavoring to draw general lessons about the cosmos from mathematical physics: They are willing to believe anything."[4] Yes, we are willing to believe anything for which there is adequate evidence. Berlinski quotes, without references, a number of scientific assertions, including one attributed to me "that the earth [*sic*] is no more significant than a single grain of sand on a vast beach." (I always capitalize Earth.) His counterargument is simply, "They are absurd; they are understood to be absurd; and what is more, assent is demanded just *because* they are absurd."[5] I think he has me confused with the ancient Christian apologist Tertullian (c. 220), who is said to have argued, "I believe because it is absurd."[6]

If Berlinski had given the location of where I supposedly said that Earth is no more significant than a single grain of sand I could look it up and tell you what I really said. However, the claim that scientists make absurd statements and then demand assent just *because* they are absurd is itself absurd. The statements made by scientists are based on objective observations and self-consistent theoretical arguments. They are independently tested and must be replicated by other scientists before being accepted into the ranks of scientific knowledge.

Becky Garrison disputes that science can say anything about God: "Just how does one use earthly empirical standards of weights, measurements, and mathematical calculations to analyze God, who transcends space, time, and matter?"[7] This is a common argument of the anti-atheists that is naively supported by many scientists. Dinesh D'Souza quotes a statement by Douglas Erwin, a paleobiologist at the Smithsonian Institution: "One of the rules of science is, no miracles allowed."[8] Is this supposed to mean that scientists would ignore a miracle if they saw one?

D'Souza also quotes biologist Barry Palevitz: "The supernatural is automatically off-limits as an explanation of the natural world."[9] Automatically? What if a supernatural explanation were found to work?

Unfortunately the US National Academy of Sciences has adopted this blinkered view of science and the supernatural: "Science is a way of knowing about the natural world. It is limited to explaining the natural world through natural causes. Science can say nothing about the supernatural. Whether God exists or not is a question about which science is neutral."[10] Only 7 percent of the members of the National Academy of Sciences believe in a personal God, with the remainder either nonbelievers or agnostics. The members of the academy, as well as most other nonbelieving scientists, have no urge to enter into a religious war with the majority of Americans.

In one of his last books, famed paleontologist (and atheist) Stephen Jay Gould attempted to negotiate a peace between science and religion by referring to them as two "non-overlapping magisteria" (NOMA), with science concerning itself with understanding the natural world while religion deals with issues of morality.[11]

Many reviewers pointed out that Gould was trying to redefine religion as moral philosophy. In fact, religions do more than just preach on morality. They make claims about the real world—about space, time, and matter—that are thereby open to scientific testing.

Furthermore, as my colleague Brent Meeker points out, religion has not exactly shown any significant expertise with respect to morality. It has supported slavery, the oppression of women, ethnic cleansing, serfdom, the divine right of kings, and extraction of testimony by torture. It has opposed anesthetics, lightning rods, sanitation, vaccination, eating meat on Friday, and birth control. It is very easy to give nonsupernatural reasons for preferring honesty to lying, for outlawing murder and theft. And in fact those moral principles were common in human society long before anyone had thought of the current major religions. So whatever useful moral prescriptions religion has provided are equally available without it.[12]

The National Academy and those scientists who agree with it that science has nothing to say about God are ignoring facts that are staring them in the face. Scientists from some of the top institutions in the country—Harvard University, Duke University, and the Mayo Clinic—have engaged in careful experiments on the efficacy of prayer.[13] If these experiments had yielded any positive results that were independently verified, replicated, and found to be sound to the consensus of the scientific community, then even new atheists would have to admit that this would be at least preliminary evidence for the existence of God warranting further study.

So far, the experiments have found no evidence that prayer works. But the point is, they might have, which sufficiently demonstrates that, no matter what National Academy or anyone else says, science is fully capable of detecting the existence of a God who acts in the lives of humans in an important way such as listening to and answering prayers.

However, theologian John Haught still thinks God is too big for puny science to detect: "Any deity whose existence could be decided by something as 'cheap' as evidence... could never command anyone's worship."[14] And, in a similar vein, "Do our new atheists seriously believe... that if a personal God of infinite beauty and unbounded love actually exists, the 'evidence' for this God's existence could be gathered as cheaply as the evidence for a scientific hypothesis?"[15]

He calls six billion dollars to build the Large Hadron Collider cheap?

You can bet Haught would change his tune if real evidence such as the ability of prayer to heal were scientifically verified. Just imagine the news stories: "Theologian John Haught welcomes evidence that God exists."

But we haven't found such evidence and so Haught must try to find ways to explain why we haven't—other than the obvious one that God does not exist. That's called "apologetics." Christians have a lot to apologize for.

Haught raises again the question of how science is legitimized: "Exactly what are the independent scientific experiments, we might ask, that could provide 'evidence' for the hypothesis that all true knowledge must be based on the paradigm of scientific inquiry?"[16]

But Haught does not refer to any specific scientist or philosopher who has made that hypothesis. As I have already mentioned and will undoubtedly do so again to drive the point home, most atheist authors fully accept and have written about the virtues of nonscientific modes of human experience such as art and music. Science is just a particularly valuable method humans have developed to learn about and partially control the physical world.

IS SCIENCE BASED ON FAITH?

Haught repeats the assertion we heard before that the new atheists have an unjustified *faith* that the real world is rational. What's the alternative, an irrational world? It's not the world that is or isn't rational. It's human beings. Being rational just means that when you talk about some subject, the words you use are well defined and the statements you make are self-consistent. How can irrational thinking with ill-defined words and inconsistent statements lead us to any credible knowledge?

The disagreement here rests on the different way scientists and intellectual theists view the world. To a scientist, calling the world "rational" or "irrational" makes no sense. It's like calling the world "hungry" or "angry." These are human mental states. Theists, on the other hand, hold to a concept of reason that is more platonic, more personal, more akin to a mystical light that suffuses the universe. In this they adhere to a more archaic idea of reason, or at least one that has not advanced along with the advance of science.[17]

Science makes no assumption about the real world being "rational." It simply applies rational methods in taking and analyzing data, following certain rules to ensure that data are as free from error as possible, and checking the logic of our models to make sure they are self-consistent. The only alternative is irrationality—error-filled data and inconsistent models.

We keep hearing that science is no less based on faith than religion. Physicist, prolific author, and Templeton Prize winner Paul Davies caused quite a stir among his fellow scientists when he wrote in an op-ed piece for the *New York Times* in 2007, "Science has its own faith-based belief system." He explains, "All science proceeds on the assumption that nature is ordered in a rational and intelligible way."[18]

This was greeted by many letters to the editor that pointed out, as I have, that our confidence in science is based on its practical success, not some logical deduction derived from dubious metaphysical assumptions.

But, given that science has not found God, Haught needs to find other justifications for belief: "If God exists, then interpersonal experience, not the impersonal objectivity of science, would be essential to knowledge of this God."[19]

Why? Science deals with observations. Interpersonal experiences should have observable consequences that science could evaluate. In a later chapter I will discuss the types of religious experiences people claim give them knowledge of God and show how these are amenable to scientific testing.

Haught argues evolution can't explain cognition: "In order to justify our cognitional confidence, something in addition to evolution must be going on during the gradual emergence of mind in natural history."[20]

This is purely a personal observation, an example of what Dawkins calls the "argument from ignorance." Just because Haught cannot understand how cognition develops naturally doesn't prove it has to be supernatural. Why can't evolution do it? Considerable work is under way in areas that connect cognitive neuroscience and evolution. While this is not yet a fully mature science, no reason has turned up to rule out the possibility that evolution can be used to explain cognition.

CAN WE TRUST OUR MINDS?

We next move to the theological argument that we can trust our minds to tell us the truth about God:

> Theology can provide a very good answer to why we can trust our minds. We can trust them because, prior to any process of reasoning or empirical inquiry, each of us simply by virtue of being or existing, is already encompassed by infinite Being, Meaning, Truth, Goodness, Beauty.[21]

Hitler trusted his mind, which told him that the Jews must be exterminated. The new atheists do not trust any minds, including their own. That's why we require the objective methods of science and reason. And how is Haught getting his values? By the same process of thought—reasoning. He certainly does not take the Bible literally when it mandates you stone to death disobedient children or the revelations of Joseph Smith when he says God wants men to have as many wives as they wish.

Haught admits that theism has no basis but blind faith to propose that our minds have access to some deeper dimension of reality:

> Theology, unlike scientism, wagers that we can contact the deepest truths only by relaxing the will to control and allowing ourselves to be grasped by a deeper dimension of reality than ordinary experience or science can by itself. The state of allowing ourselves to be grasped and carried away by this dimension of depth is at least part of what theology means by "faith."[22]

And that's what's so dangerous about it. Faith has no checks and balances, no follow-up investigation to see if an intuition works. Science does. As documented fully in the new atheist and other recent literature, it is precisely the certainty, indeed the madness, of faith—the unbridled conviction that one is doing God's work—that has over the centuries and down to the present day enabled otherwise normal human beings to commit the most cruel atrocities against their fellow humans.

DO SCIENCE AND RELIGION CONFLICT?

Theists like to point out that modern science arose in the Christian culture of Europe and that Galileo Galilei (d. 1642), Isaac Newton (d. 1727), and many other great scientists have been believers. Of course, as the cases of Galileo and Giordano Bruno (d. 1600) show, anyone living at that time didn't have much choice.

The ancient war between science and religion was chronicled in the nineteenth century by two books: *History of the Conflict between Science and Religion* (1873) by J. W. Draper,[23] and *History of the Warfare of Science with Theology* (1896) by Andrew Dickson White.[24] Since then there have been many attempts to minimize and even eliminate the claimed conflict.[25] Let me give some examples for why the conflict remains.

We saw earlier how Stephen Jay Gould sought to place science and religion into the separate compartments he called non-overlapping magisteria (NOMA). Many scientists today, believers and unbelievers alike, have adopted this view. It enables the believing scientist to switch off the science part of his brain when he goes to church. The unbelieving scientist is happy to avoid thinking about religion in going about her daily work. However, religions make testable claims about the world and science studies moral behavior. These facts prove Gould's observation to be fallacious.

Another approach called *complementarity* has been promoted in recent years by the John Templeton Foundation, founded in 1987 and funded by the (now deceased) billionaire financier John Templeton. The foundation seeks to find a common ground between science and religion. With assets of over a billion dollars, Templeton typically spends 60 million dollars each year on research grants, conferences, publications, and an annual prize of over a million dollars, designed to pay more than the Nobel Prize to a scholar who has made a major contribution to reconciling religion and science. Paul Davies was the 1996 recipient. Not all winners, such as Billy Graham and Mother Teresa (the first winner, in 1973), have had much to do with science. But several of the scientist theologians we encounter in these books have won the Templeton Prize.

I must confess I have personally received some small amounts of Templeton money for writing and debating, but Sir John would roll over in his grave if a new atheist ever won his prime award.

Although the US National Academy of Sciences and other science organizations try to keep a distance between science and religion in order to avoid conflict with the majority of ordinary citizens who are believers, the fact remains that major differences exist in the worldviews of scientists and most believers.

No place is this more evident than on the subject of evolution. The 2006 Pew survey mentioned earlier found that 65 percent of white evangelicals believe that humans and other living things have always existed in their present forms. While 62 percent of white mainline Protestants accept that life evolved over time, only 31 percent accept that the mechanism for evolution is natural selection, while 26 percent believe that a supreme being guided the process. Natural selection forms the basis of the Darwin/Wallace theory of evolution as it is applied in fields such as zoology, botany, and medicine.

The position of the Catholic Church is ambiguous on evolution. While the Church has officially acknowledged that life may have evolved, it does not extend that to the mind, which it associates with an immaterial soul. It does seem to accept natural selection, although not without God playing some role. Nevertheless, 33 percent of Catholics still think that humans and other living things did not evolve. While 59 percent accept that life evolved over time, only 25 percent attribute the process to natural selection.[26] That is, lay Catholics are not quite ready to go as far as the Church authorities, including two popes, in support of a very limited form of evolution.

In short, less than one-third of all US Christians, Catholic or Protestant, accept the theory of evolution that is universally endorsed by virtually all working biologists and all major scientific societies, including the National Academy of Sciences.

A future chapter will be devoted to the modern scientific view of the origin of the universe and its laws. We will see that this view sharply contradicts that of religion, which insists on a supernatural creation.

Some of the most prominent scientific figures of the last century have acknowledged their strong atheism: Steven Weinberg, Stephen Hawking, Steven Pinker, and the late Stephen Jay Gould, just to mention the Steves. Others living and dead include James Watson, Francis Crick, Carl Sagan, Richard Feynman, Edward O. Wilson, and, despite tales to the contrary, Albert Einstein. In a 1998 paper in *Nature*, Edward Larson reported on a survey of the 517 members of the National Academy of Sciences that found only 7 percent believe in a "personal god." Of the remainder, 72.2 percent disbelieved in a personal God and 20.8 percent expressed "doubt or agnosticism."[27]

CAN ONE BALANCE SCIENCE AND BELIEF?

Most scientists do not believe in God for two simple reasons: (1) they see no evidence for him and/or (2) they find no need to include supernatural elements in the models they build to describe their observations. Still, many are largely content to assign religion and science to separate domains, Stephen Jay Gould's "non-overlapping magisteria." This allows nonbelieving scientists to dodge the topic of religion altogether, thus avoiding jeopardizing what is most precious to them—their funding. It also permits scientists who do believe in God to compartmentalize their thinking and never discuss science and religion in the same breath.

However, this is not possible if you are Francis Collins—an evangelical Christian who headed the Human Genome Project until August 1, 2008. His 2006 best seller, *The Language of God: A Scientist Presents Evidence for Belief*, is a personal attempt to explain how he reconciles his science and his faith.[28] He describes giving a talk to a gathering of Christian physicians, recalling how warm smiles abounded when he told of his joy at being both a scientist and a follower of Christ. But when he got down to the essence of his science and tried to explain how evolution may have been God's "elegant plan" for creating humankind, the warmth evaporated from the room and some of the audience left, "shaking their heads in dismay."

Collins is fully aware of the reason the majority of Christians in the United States remain unconvinced of evolution: Darwinism implies that humanity developed by accident, contradicting the traditional teaching that humans are special, created in God's image. His explanation is that God is "outside of nature" and so knows every detail of the future. Thus, while evolution appears to us to be driven by chance, from God's perspective the outcome is "entirely specified." However, Collins does not say where that leaves human free will, which his argument implies is just an illusion.

Collins does a nice job of explaining his own work on genomes and DNA, saying that he personally finds it all the more awe-inspiring as the work of God. He also clearly points out the flaws in creationism and the more recent form of creationism called intelligent design.

However, the promise in the book's subtitle that the author will present "evidence for belief" is delivered at best inconsistently. While he admits that evolution makes the argument that humanity was designed much more complicated, he insists that God "could have" done it this way. But he fails to carry this line to its logical conclusion: God was simply not needed to do it this way. Nature could have done it all. Early on, Collins affirms what we saw above has become the disingenuous position of many scientists in the United States and of organizations such as the US National Academy of Sciences—that science has nothing to say about God and the supernatural. As I noted, this flies in the face of the facts that many reputable scientists are doing research that could, in principle, demonstrate the existence of the supernatural, such as the study of intercessory prayer.

Collins then brings up the big bang, saying: "I cannot see how nature could have created itself. Only a supernatural force that is outside of space and time could have done this." The fact that he "cannot see how" is hardly evidence for a supernatural creation. Here is yet another theist relying on the argument from ignorance.

Collins fails to account for the latest work of cosmologists who no longer view space and time as having necessarily originated with the big bang. None of the many scenarios for a natural origin of the universe that have been published in reputable journals by first-rate physicists are mentioned.

Next, Collins discusses the claim that the constants of physics are fine-tuned for life. As the philosopher David Hume pointed out centuries ago: "There can be no demonstrative argument to prove that those instances in which we have no experience resemble those of which we have had experience." In other words, we cannot use our experience in this universe, with its laws and constants, to infer what is possible in another universe with different laws and constants. The universe is not fine-tuned for life; life is fine-tuned to the universe.[29]

Furthermore, why would a perfect God make a universe so uncongenial to life that he would have to then turn around and fine-tune it? Earth-like planets should be everywhere. Or, if he wanted to, God could have designed life so it could survive anywhere, even in the vacuum of space. In fact, this is the basic flaw with all design arguments, including intelligent design in biology. The universe and life do not look at all designed; they look just as they would be expected to look if they were not designed at all.

But then, Collins himself did not become a believer because of scientific arguments or evidence. He tells us that his conversion from atheism to Christianity came from essentially one source—the writings of C. S. Lewis. His primary piece of evidence for God is not scientific—rather, he claims that "hiding in his own heart" is the clarifying principle of "Moral Law," which accounts for human altruism. Again Collins uses the "I cannot see how" argument. Turning a blind eye to the vast literature proposing an evolutionary origin for morality in humans, Collins insists that such a force can only arise from outside the material world because he cannot imagine how it could be otherwise.

No doubt some believers reading this book will be reassured that a prominent scientist is able, in his own blinkered mind at least, to reconcile science—especially evolution—with Christian belief. But it is a weak effort. If the author wished to make any significant scientific and theological statements, he would have done better to refer to the latest literature on cosmology and evolutionary psychology, and to consult theological sources besides an author of children's literature. While a favorite among evangelical Christians, Lewis is not highly regarded today by either theologians or philosophers.[30]

Geneticist Jerry Coyne has written a review, "Seeing and Believing: The Never-Ending Attempt to Reconcile Science and Religion, and Why It Is Doomed to Fail," of two more recent books by scientists who are devout Christians and argue that science and religion are compatible.[31] The books are *Saving Darwin: How to Be a Christian and Believe in Evolution* by physicist Karl W. Giberson[32] and *Only a Theory: Evolution and the Battle for America's Soul* by evolutionary biologist Kenneth R. Miller.[33]

Coyne notes how Dawkins and other new atheist writers have been denounced for "not grappling with every subtle theological argument for the existence of God." He says they miss the point: "The reason that many liberal theologians see religion and evolution as harmonious is that they espouse a theology not only alien but unrecognizable as religion to most Americans."[34]

A revealing look at the sharp division that exists among scientists on the compatibility of religion and science can be found on the Edge Web site, where comments on Coyne's reviews by several prominent scientists have been collected.[35]

Let me give one example that illuminates the differences between new and old atheism. On the Web site biologist Sir Patrick Bateson writes:

> It seems staggeringly insensitive to tell such people [believers] that they are fooling themselves and that, since they only have one life, they should get out there [and they] should enjoy it. No amount of science is going to help them to perceive the world in a way that is helpful to them. Science can be applied to relieving the conditions that oppress them—but that is a different matter. Telling them to be rational will only compound their misery.

Sam Harris responds, in a somewhat sarcastic manner:

> Patrick Bateson tells us that it is "staggeringly insensitive" to undermine the religious beliefs of people who find these beliefs consoling. I agree completely. For instance: it is now becoming a common practice in Afghanistan and Pakistan to blind and disfigure little girls with acid for the crime of going to school. When I was a neo-fundamentalist rational neo-atheist I used to criticize such behavior as an especially shameful sign of religious stupidity. I now realize—belatedly and to my great chagrin—that I knew nothing of the pain that a pious Muslim man might feel at the

sight of young women learning to read. Who am I to criticize the public expression of his faith? Bateson is right. Clearly a belief in the inerrancy of the holy Qur'an is indispensable for these beleaguered people.

CAN SCIENCE DISPROVE GOD'S EXISTENCE?

We often hear that one cannot prove or disprove the existence of God. This depends, like all logical statements, on the definition of terms. What is the definition of proof? What is the definition of God? I won't get too pedantic and ask for the definition of existence. We all have a pretty good idea what that means. Horses exist. Unicorns don't.

But what we mean by *proof* and *God* need clear specifications. In logic and mathematics we make deductive proofs in which we start with some set of assumptions and then deduce the consequences of these assumptions following procedures that have been carefully laid out and agreed to by a consensus of scholars in the fields of philosophy and mathematics. If these procedures are carried out correctly in a given proof, then the conclusion is absolutely true given that the assumptions are true.

Note that this process actually cannot tell you something that is not already embedded in the assumptions. So it is correct that you cannot use logical deduction (or mathematical deduction for that matter) to prove that God exists or does not exist. The best you can do is show that some set of assumptions about God is logically coherent or incoherent. For thousands of years, philosophers and theologians have offered logical proofs of God's existence, all of which do nothing more than establish the consistency of certain presumed attributes of God.

Still, this is a useful exercise. For example, you can prove by logical deduction that an omniscient, omnibenevolent, and omnipotent God does not exist given the gratuitous suffering in the world. By the definition of omniscience, God knows every place where there is suffering. By the definition of omnibenevolence, God does not want anything to suffer. By the definition of omnipotence, God has the capacity to undo suffering. Suffering exists, therefore an omniscient, omnibenevolent, omnipotent God does not exist.

For a series of logical disproofs of the existence of a god with various attributes, see *The Impossibility of God*, an anthology edited by Michael

Martin and Ricki Monnier.[36] Keep in mind that they only apply to the gods with those attributes carefully defined by the authors.

For an informal discussion of the logical arguments for God's existence and why they "just don't add up," see the highly entertaining book *Irreligion* by mathematician John Allen Paulos.[37] Here's how Paulos refutes the famous "first cause" argument that everything that exists has a cause and the first cause is God:

> If everything has a cause, then God does, too, and there is no first cause. And if something doesn't have a cause, it might as well be the physical world as God or a tortoise.
>
> If someone who asserts that God is the uncaused first cause (and then preens as if he's really explained something), we should thus inquire, "Why cannot the physical world itself be taken to be the uncaused first cause? After all, the venerable principle of Occam's razor advises us to 'shave off' unnecessary assumptions, and taking the world itself as the uncaused first cause has the great virtue of not introducing the unnecessary hypothesis of God."[38]

Another definition of proof can be found that is more often what we mean when we talk about proving something scientifically. This can sometimes be confusing, because science does engage in the logical and mathematical deduction of statements when it develops theories axiomatically. For example, in 1905 Einstein logically derived his theory of special relativity from two axioms: (1) The speed of light is the same in all reference frames, and (2) there is no measurement one can make to determine the absolute velocity of an object (Galileo's principle of relativity).

However, this did not "prove" that special relativity had anything to do with the natural world. That took experiments and observations. Today, after over a century of experiments and observations that agree with Einstein's theory and disagree with the earlier Newtonian predictions, we can say that special relativity is "proved beyond a reasonable doubt."

Proof beyond a reasonable doubt implies that the proof is not 100 percent certain, as in the logical or the mathematical proof. On the other hand, it goes further than these proofs in telling us something that is not already built into the assumptions. And this is a point so often missed by theists when they argue that science is no better than theology—scientific proof can tell us something we do not already know. Theology, unless based on empirical evidence, does not.

The criterion of scientific proof is similar to that used in courtrooms, where decisions are made "beyond a reasonable doubt." Unlike court cases, however, science places no limits on appeals to judgments. No "Supreme Court" exists as the final arbiter. No matter how long it takes, hundreds of years in the case of Newtonian mechanics, a theory can be supplanted by a new theory when the new theory does a better job of agreeing with the data.

Note that I did not say that special relativity proved Newtonian mechanics "wrong." The observed differences between the Einsteinian and Newtonian predictions are mostly only evident at speeds near the speed of light. Physical theories are just models that we put to use in helping us classify and predict observations. Newtonian mechanics remains perfectly useful and "true" at low relative velocities. It's what most students learn when they take physics, and what most professionals, such as engineers, use when they use physics.

In any case, when I say that science has proved that God does not exist I mean that a god with specific properties (whom I identify as God with a capital G) can be disproved scientifically beyond a reasonable doubt. We have already seen that science could prove any god's existence beyond a reasonable doubt, simply by observing some phenomenon that we can rule out beyond a reasonable doubt as being natural and we can associate with that god. One example I gave was the success of a specific type of intercessory prayer associated with one religion and no others. Another example was a revealed truth that can be verified scientifically and demonstrated to be unknowable by natural means, such as the successful prediction of some future event.

Of course, we have not seen such proofs beyond a reasonable doubt of God's existence. On the other hand, I believe that a good case can be made that science can now state, beyond a reasonable doubt, that the Abrahamic God does not exist.

While theologians might quibble about the attributes of God and the differences that exist between the Abrahamic religions and their various sects, it is beyond argument that the God worshipped and prayed to by Jews, Christians, and Muslims is an active participant in the world and in human lives. If we are to hypothesize the existence of such a God, we can then infer certain observations that should be detectable, if not to the naked eye in humans, then in the careful, objective analysis of data from the highly sensitive instruments of science.

In my 2007 book, *God: The Failed Hypothesis—How Science Shows That God Does Not Exist,* I made such a hypothesis and analyzed the observations that should have been made to verify that hypothesis. I will not repeat this analysis here. Suffice it to say that even the most devout believer must admit that science has not discovered any kind of God. My argument is simply that science, or even the human senses, should by now have turned up empirical evidence for a moral God who acts in the universe in any significant way.

I qualified my statement above to refer only to a "moral" God because we can hypothesize an immoral god who deliberately hides even from humans who are open to belief (see discussion in chapter 2).[39]

IS ANY GOD CONSISTENT WITH SCIENCE?

We saw in chapter 1 that 44 percent of all Americans do not believe in a god who plays an important role in the world or in their personal lives. That is, they do not believe in the traditional Christian God. Rather than theism, their beliefs should be more accurately classified as *deism.*

Deism originated during the eighteenth-century Age of Enlightenment. Let me refer to that form of deism specifically as *Enlightenment deism* because a viable modern deism is forced by scientific developments to be something different. This was the first time in Western Christendom that educated people trusted reason over faith. Many of the most prominent of the Founding Fathers of the American republic were deists, notably Thomas Paine, Benjamin Franklin, and probably our first four presidents. The "creator" whom Thomas Jefferson wrote into the Declaration of Independence was the deist God.

It made no sense to these rational thinkers that an all-powerful, all-knowing God would have to step in to make changes once he had set the universe in motion. In 1687 Isaac Newton had published his laws of gravity and motion, which allowed the prediction of the motion of any material body with knowledge of the initial position and velocity of the body and the force acting on it. This suggested the universe was a vast machine or clockwork, with everything predetermined. Since God made these laws, the Enlightenment deists saw no reason for him to act further.

If humans are made of material particles and nothing more, then Enlightenment deism implies that we have no free will. However, free will is still possible if we have immaterial souls that enable us to freely control our material bodies. Certainly many humans, including deists, would argue vehemently from experience that they possess free will.

Still, almost half of today's Americans are in tune with at least a version of the deism of the Founding Fathers. Perhaps deism has been around all along as an unacknowledged private feeling, the simple common sense that a perfect god who created the universe and its laws should never need to step in and make changes to events that are already predetermined.

However, we can now with a high degree of confidence rule out Enlightenment deism. The Heisenberg uncertainty principle of quantum mechanics has shown that the motions of bodies contain an element of chance. Physicists cannot in fact predict their motions with unlimited accuracy. The universe is not a vast machine in which everything that happens is fully determined by what went on before.

This still allows for the possibility of a god who built the universe and its laws but who also added a large amount of randomness. In fact, this is just the "God who plays dice" whom Einstein refused to accept.

Many theologians, including a number who have excellent scientific credentials, have approached the problem of reconciling science and religion by modifying theology so that it is consistent with science. I have written sympathetically about them in the past because they are willing to accept science and not try to change it to suit their beliefs.[40]

Modern Christian theologians see the need to address the question of God's action in the universe and have held a series of conferences on the subject at the Vatican. Several physics-trained theologians have suggested that quantum mechanics and chaos theory provide possible mechanisms. However, as I discuss in detail in *Quantum Gods*, these mechanisms still require God to violate the laws of physics when he steps in to change the motions of bodies.

Other theologians and theistic scientists have found almost no choice but to assert some version of a deist god without identifying it as such, the God who plays dice discussed above. In *God, Chance and Purpose*, statistics scholar David Bartholomew elevates chance to be an integral part of God's creation, the means by which God achieves creativity in the world.[41] This is still a deist god since the dice are tossed before the creation. So far,

Bartholomew and the other "deologians," if I may invent a word, have not been able to reconcile such a god with the Christian God who, according to all Christian teachings, acts in the everyday world and must be worshipped and prayed to so that his actions are favorable.

NOTES

1. David Marshall, *The Truth behind the New Atheism* (Eugene, OR: Harvest House, 2007), p. 209.

2. See, for example, Richard Dawkins, *Unweaving the Rainbow* (Boston: Houghton Mifflin, 1998).

3. David Berlinski, *The Devil's' Delusion* (New York: Crown Forum, 2008), p. 6.

4. Ibid., p. 4.

5. Ibid., pp. 8–9.

6. See Wikipedia, http://en.wikipedia.org/wiki/Credo_quia_absurdum (accessed October 15, 2008).

7. Becky Garrison, *The New Atheist Crusaders and Their Unholy Grail* (Nashville: Thomas Nelson, 2007), p. 43.

8. Dinesh D'Souza, *What's So Great about Christianity?* (Washington, DC: Regnery, 2007), p. 157. Originally cited by Kenneth Chang, "In Explaining Life's Complexity, Darwinists and Doubters Clash," *New York Times*, August 22, 2005.

9. Barry Palevitz, "Science vs. Religion" in *Science and Religion: Are They Compatible?* ed. Paul Kurtz (Amherst, NY: Prometheus Books, 2003), p. 175.

10. National Academy of Sciences, *Teaching about Evolution and the Nature of Science* (Washington, DC: National Academy of Sciences, 1998), p. 58. Online at http://www.nap.edu/catalog/5787.html (accessed March 5, 2006).

11. Stephen Jay Gould, *Rocks of Ages: Science and Religion in the Fullness of Life*, Library of Contemporary Thought (New York: Ballantine, 1999).

12. Brent Meeker, private communication.

13. H. Benson, J. A. Dusek, J. B. Sherwood, P. Lam, C. F. Bethea, et al., "Study of the Therapeutic Effects of Intercessory Prayer (STEP) in Cardiac Bypass Patients: A Multicenter Randomized Trial of Uncertainty and Certainty of Receiving Intercessory Prayer," *American Heart Journal* 151, no. 4:934–42.

14. John F. Haught, *God and the New Atheism* (Louisville: Westminister John Knox Press, 2008), p. 44.

15. Ibid., p. 46.

16. Ibid., p. 45.

17. Taner Edis, "Is the Universe Rational?" http://www2.truman.edu/~edis/writings/articles/RationalUniverse.html (accessed February 21, 2009).

18. Paul Davies, "Taking Science on Faith," *New York Times*, November 24, 2007.

19. Haught, *God and the New Atheism*, p. 46.

20. Ibid., p. 49.

21. Ibid., p. 50.

22. Ibid., p. 60.

23. John William Draper, *History of the Conflict between Religion and Science* (London: Watts & Co., 1873).

24. Andrew Dickson White, *A History of the Warfare of Science with Theology in Christendom* (New York: D. Appleton & Co., 1896).

25. Kurtz, ed., *Science and Religion: Are They Compatible?*

26. Pew Research Center, "Many Americans Uneasy with Mix of Religion and Politics," http://pewforum.org/publications/surveys/religion-politics-06.pdf (accessed November 7, 2008).

27. Edward J. Larson, "Leading Scientists Still Reject God," *Nature* 294, no. 6691 (1998): 313.

28. Francis S. Collins, *The Language of God: A Scientist Presents Evidence for Belief* (New York: Free Press, 2006).

29. The above discussion was based on the author's review of *The Language of God* that appeared in Victor J. Stenger, "A Lack of Evidence," *Physics World* 19, no. 10 (2006): 45–46.

30. John Beversluis, *C. S. Lewis and the Search for Rational Religion* (Amherst, NY: Prometheus Books, 2007).

31. Jerry A. Coyne, "Seeing and Believing: The Never-Ending Attempt to Reconcile Science and Religion, and Why It Is Doomed to Fail," *New Republic*, February 4, 2009.

32. Karl Giberson, *Saving Darwin: How to be a Christian and Believe in Evolution* (New York: HarperOne, 2008).

33. Kenneth R. Miller, *Only a Theory: Evolution and the Battle for America's Soul* (New York: Viking Penguin, 2008).

34. Coyne, "Seeing and Believing."

35. John Brockman, "Edge: The Reality Club," http://www.edge.org/3rd_culture/coyne09/coyne09_index.html, 2009 (accessed April 8, 2009).

36. Michael Martin and Ricki Monnier, *The Impossibility of God* (Amherst, NY: Prometheus Books, 2003).

37. John Allen Paulos, *Irreligion: A Mathematician Explains Why the Arguments for God Just Don't Add Up* (New York: Hill and Wang, 2008).

38. Ibid., pp. 4–5.

39. J. L. Schellenberg, *Divine Hiddenness and Human Reason* (Ithaca, NY: Cornell University Press, 1993); Theodore M. Drange, *Nonbelief & Evil: Two Arguments for the Nonexistence of God* (Amherst, NY: Prometheus Books, 1998).

40. Victor J. Stenger, *Has Science Found God? The Latest Results in the Search for Purpose in the Universe* (Amherst, NY: Prometheus Books, 2003), pp. 307–38; Victor J. Stenger, *God: The Failed Hypothesis—How Science Shows That God Does Not Exist* (Amherst, NY: Prometheus Books, 2008), pp. 235–37; Victor J. Stenger, *Quantum Gods: Creation, Chaos and the Search for Cosmic Consciousness* (Amherst, NY: Prometheus Books, 2009).

41. David J. Bartholomew, *God, Chance, and Purpose: Can God Have It Both Ways?* (Cambridge: Cambridge University Press, 2008).

4. THE DESIGN DELUSION

> **We are trying to understand how we have got a complicated world, and we have an explanation in terms of a slightly simpler world, and we explain that in terms of a slightly simpler world and it all hangs together down to an ultimately simple world. Now, God is not an explanation of that kind. God himself cannot be simple if he has power to do all the things he is supposed to do.**
>
> —Richard Dawkins

DESIGNED FOR LIFE?

For most theists, one of the strongest arguments for the existence of God is the apparent design that they imagine they see in the universe and in living things. This *argument from design* is ancient, but it has surfaced in recent years in two forms that are scientific rather than philosophical, and so these are inherently more capable of reaching a definitive conclusion since they are based on data and not reasoning from arguable premises. The new forms of the argument from design are:

1. The argument from the evidence for *fine-tuning* of the universe (better known by the misleading designation, the *anthropic principle*).
2. The argument from the evidence for *intelligent design* in the complex structures of living organisms.

Although the second has received greater public attention, most science-savvy theologians agree with most scientists that intelligent design, at least as it has been formulated so far, is a failure. Theologians are far more impressed by the fine-tuning argument and they have received support from a number of prominent scientists who profess not to be believers but admit that the facts are puzzling and require explanation.

On the level of physics and cosmology, theists claim that the universe is so finely tuned for life that any slight difference in the physical parameters and life would not have been possible. They conclude that this provides overwhelming evidence that the universe was designed for life, possibly even human life, and therefore a designer must exist.

Although those making this argument quickly jump to the conclusion that the designer is the particular god they worship, he, she, or it could just as well be the deist god or some impersonal supernatural power such as the "Force" in *Star Wars*. Or, perhaps the designer is a natural power somehow within the material universe.

I have addressed the fine-tuning arguments in a number of books and articles.[1] However, the subject continues to develop, as do my responses, and I will not be too repetitive here. I will discuss several important examples that I have not covered in these earlier writings.

A LOGICAL ARGUMENT

But before getting to the data, let me present the typical logical argument for fine-tuning and show how it fails. Remember, logic cannot be used to tell us something not already embedded in the postulates we assume when we begin the deductive process. However, it can be used to tell us that something is wrong because the postulates are unjustified or the argument is logically inconsistent. Let me play philosopher for a moment and try to state the fine-tuning argument as a (not too pedantic) philosopher might:

1. Postulates

- There exists only one universe, the one we live in.
- The physical parameters of our universe were not all fixed by some fundamental physical law.
- A huge variation in the ranges of the unfixed parameters was possible, making any given set highly unlikely by chance.
- Our universe possesses just those physical parameters that make life in our universe possible.
- Humans and other earthly life would not exist if our universe had slightly different parameters.

2. Conclusions

- If the universe did not have the parameters it does, we would not be here to talk about it. This is the *weak anthropic principle* and is a simple tautology.[2]
- Because the set of parameters of our universe is so unlikely to have occurred by chance, it most probably occurred by design, with humanity part of that design.

3. Refutations

- Not yet having a final theory of everything (TOE), we cannot know that more than one set of parameters is possible. Indeed, current string theory suggests on the order of 10^{500} different possible sets of parameters.[3]
- Since our only experience is with our universe and its parameters, we have no way of knowing that a huge range of parameters is possible or anything about the distribution of those parameters so that any given set is unlikely by chance.
- Since our only experience is with our universe and its parameters, we have no way of knowing what sets of parameters might still lead to some form of life different from ours but just as complex and still containing intelligent beings.
- We have no reason to assume there aren't many universes. Even if our particular universe is highly unlikely, the chance that we are one of many could be as high as 100 percent.

The multiple universe, or *multiverse*, scenario is severely criticized by theists as unobservable. Of course God is unobservable, too, so the best the theists can claim is a standoff. Actually, there are already some ideas about how we might detect the presence of other universes, but they are highly speculative and too technical to mention here. Less speculative are the theories that predict the multiverse, since these are part of science. Scientists theorize about the unobservable all the time, such as quarks and dark matter, and there is nothing wrong with that as long as the science being used is well established.

In fact, we have good reasons to believe in the multiverse. As physicist Aurélian Barrau writes in his review of multiverse physics, "The idea of multiple universes is more than a fantastic invention. It appears naturally within several theories and deserves to be taken seriously."[4] I strongly recommend this article because it will demonstrate that the multiverse is not just some vague atheist notion dreamed up to counter the fine-tuning argument. It is sober, mainstream physics.

In conclusion, the fine-tuning argument can be countered logically. But all I have proved is that the formulation of the fine-tuning argument given above fails because several of the postulates cannot be justified. They might still be true, or some other formulation might be true. You just cannot prove it's true.

Let us take a different tack and assume the postulates are true and look at some of the scientific facts that lead science-savvy theists and some nontheistic scientists to still think that something funny is going on that they don't want to dismiss, as many more hard-nosed types do, by just saying, "Well, that's the way it is."

SCIENTIFIC ARGUMENTS

A list of thirty-four parameters that seem to be fine-tuned has been assembled by Rich Deem on the God and Science Web site.[5] His main reference is physicist and Christian apologist Hugh Ross and his popular book *The Creator and the Cosmos*, first published in 1993.[6] Ross is the founder of Reasons to Believe, which is a self-described "international and interdenominational science-faith think tank *providing powerful new reasons from science to believe in Jesus Christ*."[7] A long list of claimed "design evidences" can be found on its Web site.[8]

Several of Deem's and Ross's constants, such as the speed of light in a vacuum, *c*, Newton's constant of gravity, *G*, and Planck's constant, *h*, are just arbitrary numbers that are determined simply by the unit system you are using. They can be set equal to any number you want, except zero, with no impact on the physics. So no fine-tuning can possibly be involved, just as the number *pi* is not fine-tuned.

Deem does not actually mention *h* explicitly, but it comes in when he talks about the "magnitude of the Heisenberg uncertainty principle" being fine-tuned so that oxygen transport to the cells is just right. The magnitude of the uncertainty principle is simply Planck's constant, *h* (technically, *h* divided by 4π).[9] We can safely assume that life evolved in such a way that the energy transport was just right, adjusting to the physical parameters such as they are.

I will focus on the five parameters that have the most significance because, if interpreted correctly, they pretty much rule out almost any conceivable kind of life without fine-tuning. Copying, with minor modifications, the table from Deem:

Table 4.1. Five parameters that seem to be the most highly fine-tuned for life to exist

Parameter	*Max. deviation*
Ratio of electron to protons	1 part in 10^{37}
Ratio of electromagnetic force to gravity	1 part in 10^{40}
Expansion rate of the universe	1 part in 10^{55}
Mass density of the universe*	1 part in 10^{59}
Cosmological constant	1 part in 10^{120}

*Deem says mass here, but based on his further discussion, I infer he meant mass *density*.

Deem does not give any references to scientific papers showing calculations for the "maximum deviations" listed in the table. However, I will admit that the features a universe would have for slightly different values of these parameters, all other parameters remaining the same, would render our form of life impossible. Indeed, this extends to any form of life even remotely like ours, that is, one that is based on a lengthy process, chemical or otherwise, by which complex matter evolved from simpler matter. This is the main reason these parameters are significant. As we will see below, in all the other examples people give, *some* form of life is still possible, just not our form specifically.

Let me discuss each in turn. Note that the arguments all apply to our universe where we assume that none of the laws of physics are different. All that is different is the value of some of the numbers we put into those laws, so we are not making any assumption about other universes.

Ratio of Electrons to Protons

The claim is that if the ratio of the numbers of electrons to protons were larger, electromagnetism would dominate over gravity and galaxies would not form. If smaller, gravity would dominate and chemical bonding would not occur. This assumes the ratio is some arbitrary constant. In fact, the number of electrons equals exactly the number of protons for a very simple reason: the universe is electrically neutral—so the two, having opposite charges, must balance.

Here is a clear but slightly technical explanation for how this all came about in the early universe from a book by astronomer Peter Schneider:

> Before pair annihilation [the time when most electrons annihilated with antielectrons, or *positrons*, producing photons] there were about as many electrons and positrons as there were photons. After annihilation *nearly* all electrons were converted into photons—but not entirely because there was a very small excess of electrons over positrons to compensate for the positive electric charge density of the protons. Therefore the number density of electrons that survive the pair annihilation is exactly the same as the number density of protons, for the Universe to remain electrically neutral.[10]

So, no fine-tuning happened here. The ratio is determined by conservation of charge, a fundamental law of physics.

Ratio of Electromagnetic Force to Gravity

The source of the huge difference in strengths between the electromagnetic and gravitational forces of a proton and an electron, $N_1 = 10^{39}$, has been a long-standing problem in physics, first being mentioned by Hermann Weyl in

1919. If they were anywhere near each other, stars would collapse long before they could provide either the materials needed for life on planets or the billions of years of stable energy needed for life to evolve. Why is $N_1 = 10^{39}$? You would expect a natural number to be on the order of magnitude of 1.

But note that N_1 is not some universal measure of the relative strengths of the electric and gravitational forces. It's just the force ratio for a proton and an electron. The proton is not even an elementary particle. It is made of quarks. For two electrons the ratio is 10^{42}. If we have two unit-charged particles, each with a mass of 1.85×10^{-9} kilograms, the two forces would be equal!

So the large value of N_1 is simply an artifact of the use of small masses in making the comparison. The force ratio is hardly "fine-tuned to 39 orders of magnitude," as you will read in the theist literature. The reason N_1 is so large is that elementary particle masses are so small. According to our current understanding of the nature of mass, elementary particles all have zero bare mass and pick up a small "effective mass" by interacting with the background Higgs field that pervades the universe. That is, their masses are "naturally" very small.

If particle masses were not so small, we would not have a long-lived, stable universe and wouldn't be here to talk about it. This is an expression of the anthropic principle. But note that I am not invoking it—I am explaining it.

Several physicists, including myself, have done computer simulations where we generate universes by varying *all* the relevant parameters. In my case I randomly varied the electromagnetic force strength, the mass of the proton, and the mass of the electron by ten orders of magnitude around their existing values in our universe.[11] The gravitational force strength was fixed. That is, I allowed the ratio of forces to vary from 10^{34} to 10^{44}! These are a long way from 10^{39}. I found that over half of the universes generated had stars with lifetimes of at least ten billion years, long enough for life of some kind to evolve.[12]

In a more recent study I have also varied the strength of the strong nuclear force and placed further limitations on the characteristics of the generated universes. For example, I ensure that atoms are much bigger than nuclei and have much lower binding energy. I have demanded that planetary days be at least ten hours long and stars be far more massive than planets. I find that 20 percent of the universes have these properties.

My study was rather simple. More-advanced studies, which reach the same basic conclusion, have been made by Anthony Aguire, Roni Harnik, Graham D. Kribs, Gilas Perez, and Fred C. Adams.[13]

Expansion Rate of the Universe

The fine-tuning of the expansion rate of the universe is one of the most frequent examples given by theologians and philosophers.[14] Deem says that if it were slightly larger, no galaxies would form; if it were smaller, the universe would collapse.

This has an easy answer. If the universe appeared from an earlier state of zero energy, then energy conservation would require the exact expansion rate that is observed. That is the rate determined precisely by the fact that the potential energy of gravity is exactly balanced by the kinetic energy of matter.

Let me try to explain this in detail so that, once again, it is clear that I am merely stating a simple fact of physics. Suppose we wish to send a rocket from Earth to far outside the solar system. If we fire the rocket at exactly 11.2 kilometers per second, what is called the *escape velocity* for Earth, its kinetic energy will exactly equal the negative of its gravitational potential energy, so the total energy will be zero. As the rocket moves away from Earth, the rocket gradually slows down. Its kinetic energy decreases, as does the magnitude of its potential energy, the total energy remaining constant at zero because of energy conservation. Eventually when the rocket is very far from Earth and the potential energy approaches zero, its speed relative to Earth also approaches zero.

If we fired the rocket at just under escape velocity, the rocket would slow to a stop sooner and eventually turn around to return to Earth. If we fired it at a slightly higher speed, the rocket would keep moving away and never stop.

In the case of the big bang, the bodies in the universe are all receding from one another at such a rate that they will eventually come to rest at a vast distance. That rate of expansion is very precisely set by the fact that the total energy of the system was zero at the very beginning, and energy is conserved.

So, instead of being an argument for God, the fact that the rate of expansion of the universe is exactly what we expect from an initial state of zero energy is a good argument against a creator. Once again, we have no fine-tuning because the parameter in question is determined by a conservation principle, in this case conservation of energy.

Mass Density of the Universe

If the mass density of the universe were slightly larger, then overabundance of the production of deuterium (heavy hydrogen) from the big bang would cause stars to burn too rapidly for life on planets to form. If smaller, insufficient helium from the big bang would result in a shortage of the heavy elements needed for life.

The answer is the same as the previous case. The mass density of the universe is precisely determined by the fact that the universe starts out with zero total energy.

Cosmological Constant

This is considered one of the major unanswered problems in physics. The *cosmological constant* is a term that arises in Einstein's general theory of relativity. It is basically equivalent to the energy density of empty space that results from any curvature of space. It can be positive or negative. If positive it produces a repulsive gravitational force that accelerates the expansion of the universe.

For most of the twentieth century it was assumed that the cosmological constant was identically zero, although no known law of physics specified this. At least no astronomical observations indicated otherwise. Then, in 1998, two independent groups studying supernovas in distant galaxies discovered, to their great surprise since they were looking for the opposite, that the expansion of the universe was accelerating. This result was soon confirmed by other observations, including those made with the Hubble Space Telescope.

The component of the universe responsible for the acceleration was dubbed *dark energy*. It constitutes 73 percent of the total mass of the universe. (Recall the equivalence of mass and energy given by $E = mc^2$.) The natural assumption is to attribute the acceleration to the cosmological constant, and the data, so far, seem to support that interpretation.

Theorists had earlier attempted to calculate the cosmological constant from basic quantum physics. The result they obtained was 120 orders of magnitude larger than the maximum value obtained from astronomical observations.

Now this is indeed a problem. But it certainly does not imply that the cosmological constant has been fine-tuned by 120 orders of magnitude. What it implies is that physicists have made a stupid, dumb-ass, wrong calculation that has to be the worst calculation in physics history.

Clearly the cosmological constant is small, possibly even zero. This can happen in any number of ways. If the early universe possessed, as many propose, a property called *supersymmetry*, then the cosmological constant would have been exactly at zero at that time. It can be shown that if negative energy states, already present in the calculation for the cosmological constant, are not simply ignored but counted in the energy balance, then the cosmological constant will also be identically zero.

Other sources of cosmic acceleration have been proposed, such as a field of neutral material particles pervading the universe that has been dubbed *quintessence*. This field would have to have a negative pressure, but if it is sufficiently negative it will be gravitationally repulsive.

In short, the five greatest fine-tuning proposals show no fine-tuning at all. The five parameters considered by most theologians and scientists to provide the best evidence for design can all be plausibly explained. Three just follow from conservation principles, which argue against rather than for any miraculous creation of the universe. They can be turned around and made into arguments against rather than for God.

Answers can be given for all the other parameters on Deem's list. Not a single one rules out some kind of life when the analysis allows other parameters to vary.

The same can be said about other compilations of parameters that appear at first glance to be fine-tuned. Even some of the most respected scientists have made the mistake of declaring a parameter "fine-tuned" by only asking what happens when it is varied while all other parameters remain the same. A glaring example is provided by Sir Martin Rees, the Astronomer Royal of the United Kingdom. In his popular book *Just Six Numbers*,[15] Rees claims that a quantity he calls "nuclear efficiency," ε, defined as the fraction of mass of helium that is less than two protons and two neutrons, is fine-tuned to $\varepsilon = 0.07$. If $\varepsilon = 0.06$, deuterium would be unstable because the nuclear force would not be strong enough to keep the electrical repulsion between protons from blowing it apart. If $\varepsilon = 0.08$, the nuclear force would be strong enough to bind two protons together directly and there would be no need for neutrons and to provide additional attraction. In that case there would be no deuterium or any other nuclei containing neutrons.

But this all assumes a fixed strength of the electromagnetic force, which is given by the dimensionless parameter α, which has a value of 1/137 in the existing universe. In the first case, for any value of α less than 1/160 the deuteron will be stable because the electrical repulsion will be too weak to split it apart. In the second case, for any value of α greater than 1/120 the electrical repulsion will be too great for protons to bind together without the help of neutrons, so deuterons and other neutron-rich nuclei will exist. So stable nuclei are possible for a wide range of the two parameters ε and α and neither are fine-tuned for life.

Let me mention one parameter where the answer to the claim of fine-tuning is ridiculously simple. The masses of neutrinos are supposedly fine-tuned since their gravitational effects would be too big or too small if they were different and this would adversely affect the formation of stars and galaxies. But that assumes that the number of neutrinos in the universe is fixed. It is not. It is determined by their masses. If heavier, there would be fewer. If lighter, there would be more. Whatever the masses, the gravitational effects of neutrinos would be the same.

In short, there is no scientific basis for the claim that the universe is fine-tuned for life. Indeed, the whole notion makes no sense. Why would an omnipotent god design a universe in which his most precious creation, humanity, lives on the knife-edge of extinction? This god made a vast universe that is mostly empty space and then confined humankind to a tiny speck of a planet, where it is destined for extinction long before the universe becomes inert. He could have made it possible for us to live anywhere. He also could have made it possible to live in any conceivable universe, with any values for its parameters. Instead of being an argument for the existence of god, the apparent fine-tuning of the constants of physics argues against any design in the cosmos.

EARTHLY DESIGN

Let us now move to the question of design as it is imagined by theists to be empirically evident in living things on Earth. By now the history must be well known to readers, so I will only briefly summarize that story and bring up a few new wrinkles. Although it can be traced back to antiquity, the argument from design received its most significant boost from the work of Archdeacon William Paley in *Natural Theology*, first published in 1802.[16] Paley compared a rock to a watch and noted that the watch was clearly an artifact. He then argued that the human body was an artifact like the watch.

The argument from design was originally a god of the gaps argument, when science had no explanation for how the complexity of life arose naturally. When, fifty years after Paley, in 1859, Charles Darwin and Alfred Russel Wallace showed how complex biological systems can evolve from simpler ones by means of natural selection, the argument from design evolved from a god of the gaps into an argument from ignorance. There no longer was a gap. When used today it is given in the form, "I can't see how life can have happened naturally, therefore, it must have come about supernaturally."

Evolution immediately came under attack from religious spokesmen because it clearly conflicted with Genesis, and it remains under attack today from those who read the Bible literally. Since the Catholic Church's ultimate authority is not the Bible but the pope, it has found it easier to accept the verdict of science. Some proponents of evolution, such as the late, famed paleontologist Stephen Jay Gould, see no conflict with religion. They jumped with joy when they read in 1996 that Pope John Paul II had declared that "evolution is more than a hypothesis."[17] The pope seemed to say that evolution was now an established fact. However, as it turned out, what he said was quite the opposite.

In his message of October 22, 1996, to the Pontifical Academy of Sciences, John Paul II refers to "Encyclical Humani Generis" (1950), composed by his predecessor, Pius XII, as stating that "there was no opposition between evolution and the doctrine of the faith about man and his vocation, on condition that one did not lose sight of several indisputable points (cf. AAS 42 [1950], pp. 575–576)."[18]

In his message, John Paul II actually hedged considerably on his acceptance of evolution, implying it has not yet been validated and there is more than one hypothesis in the theory. When this message first came out, the translation read "evolution is more than *a* hypothesis," which evolution enthusiasts gleefully interpreted as the pope's strong support for evolution as fact, not just theory. A full reading of the message, however, makes it clear that the Holy Father was claiming that there are "several theories" of evolution that are still under dispute, hardly the position of today's evolutionists.

Further, John Paul II elaborated on what Pius XII had called "indisputable points." These had to do with soul and mind. Again, referring to Pius XII, John Paul II says:

> If the human body takes its origin from pre-existent living matter the spiritual soul is immediately created by God. Consequently, theories of evolution which, in accordance with the philosophies inspiring them, consider the mind as emerging from the forces of living matter, or as a mere epiphenomenon of this matter, are incompatible with the truth about man. Nor are they able to ground the dignity of the person.[19]

John Paul II's successor, Benedict XVI, has moved the Church even further from the scientific consensus by asserting "the theory of evolution is not a complete, scientifically proven theory."

Mainline Protestant churches (a dying breed in the United States, a dead breed in Europe) have stated their support for evolution, but the more conservative fundamentalist, evangelical, and Pentecostal churches continue to fight it tooth and nail. Eastern Orthodox Christians, Muslims, and Orthodox Jews also reject evolution.

Any Christian, Jew, or Muslim who accepts evolution, as do most of the scientists of these faiths, must be confronted with the fact that, according to the conventional interpretation, the human species is an accident. This violates the fundamental teaching by all these faiths that humanity is special. The answer usually given is that God, in fact, guides evolution. So, these so-called evolution believers are really believers in divine design after all.

INTELLIGENT DESIGN

Intelligent design grew out of the earlier creation science movement, which attempted to show that the account of creation presented in Genesis—an Earth only a few thousand years old, the special, individual creation of each species, and a flood that inundated the world—had a scientific basis. After attempts to teach creation science in schools were declared unconstitutional in the courts, a new group of opponents of evolution developed a more sophisticated approach that made no attempt to interpret the Bible as a science text but introduce a new "science" they called *intelligent design* (ID).

The proponents of intelligent design understood the point I have made about the god of the gaps argument for the existence of God, that it is not sufficient just to have a gap in scientific knowledge; you must prove that the gap will never be filled by future scientific developments. However, they ended up using the argument from ignorance.

Perhaps the best-known proponent of ID is Lehigh University biochemist Michael Behe. In a popular book called *Darwin's Black Box*, Behe asserted that certain biological structures are "irreducibly complex," that is, they contain parts that could not have evolved without the rest of the structure since then they would not be able to function and thus required a designer outside of nature. [20] He likened this to the parts of a mousetrap that individually cannot catch a mouse or perform any other task. Other examples included bacterial flagella and blood clotting. Evolutionary biologists quickly pointed out that biological structures often change their functions during evolution and showed that all of Behe's examples were in fact reducible and easily explained by conventional natural selection.[21] Some of those explanations were already in the literature.

The other main figure in the ID movement was philosopher and mathematician William Dembski. Dembski also attempted to show that science would never be able to fill what he, like Behe, imagined was a gap in scientific knowledge. Dembski also proposed that complexity was irreducible to simplicity, but in a more general way that applied to all physical systems, not just biology. In his book *Intelligent Design* he proposed a "law of conservation of information" in which he claimed that it was impossible for any natural process to increase the information of a system. Thus, a simple, low information system cannot evolve into a more complex, higher information system without the help of an outside designer.[22]

In my 2003 book, *Has Science Found God?* I disproved Dembski's law of conservation of information.[23] His quantitative definition of information followed that of conventional informational theory in which information transfer is proportional to the change in entropy of a system. However, entropy is not a conserved quantity in physics. In fact, the second law of thermodynamics says that when two bodies interact and exchange entropy, the total entropy of the two-body system will either stay the same or increase, assuming no other bodies are interacting with it. For example, when a higher temperature body comes in contact with a lower temperature one, heat flows from the hotter to the cooler body until they reach an equilibrium temperature. During the process, the total entropy increases and information has been lost.

There are many examples in nature of information being generated naturally. Consider a magnetic needle standing vertically. It provides no information about compass direction. Then it falls over and points north. Information about the direction north has been generated naturally, with no outside designer.

THE END IN COURT

Intelligent design reached an ignominious end in Dover, Pennsylvania, in December 2000 when US federal judge John E. Jones III ruled in *Kitzmiller et al. v. Dover Area School District et al.* that a pro–intelligent design disclaimer could not be read to public school students.[24] The evidence presented at the trial showed beyond a reasonable doubt that the motives behind those who proposed the disclaimer were religious and thus represented an unconstitutional infringement by church on state. This was sufficient to abolish the disclaimer, but then the judge went too far.

I agree with my University of Colorado colleague, philosopher of physics Bradley Monton, who has written that other parts of the Jones decision were deeply flawed.[25] Basically the judge followed the precedent set in 1982 when a federal judge in Little Rock, Arkansas, ruled unconstitutional a state law that called for the "balanced treatment for Creation-Science and Evolution-Science." Again, this decision was justified based on the strong evidence of a religious motivation behind the act. However, the Arkansas judge, William Overton, went further than needed in declaring that Creation Science was not science. In this he was setting a demarcation criterion for deciding between what is and what is not science that did not reflect the consensus of scientists or philosophers of science. At the time,

Overton's decision was strongly criticized by my then University of Hawaii colleague, the prominent philosopher of science Larry Laudan.[26]

In 2005 in Dover, Judge Jones similarly ruled that intelligent design is not science. I agree with Monton (an avowed atheist, incidentally) that intelligent design is science. However, in its current form with the claims of Behe and Dembski described above, it is wrong science. A wrong science is still science, but it should not be taught in school because it is wrong, not because it isn't science. Teachers should not be teaching that the Earth is flat (as the Bible implies) or that disease is caused by an imbalance of humors. So they should not be teaching irreducible complexity or the law of conservation of information. Because they are wrong.

The intelligent design movement was bankrolled by an evangelical Christian organization called the Discovery Institute.[27] Its original stated goal was to "defeat scientific materialism and its destructive moral, cultural, and political legacies" and "renew" science and culture along evangelical Christian lines. They have called this the "wedge strategy."[28] Now that intelligent design in biology seems to be scientifically (though not yet politically) dead, the institute is beginning to pursue other areas where they might demonstrate the existence of a creator. This would not be objectionable if they did so with an open mind and were willing to accept—as science does—the verdict of the data. However, their mode of operation is to scan for data and arguments that support their faith and discard any data and arguments that do not.

At the Discovery Institute's August 2008 Insider's Briefing on Intelligent Design, two of the five speakers were neuroscientists. *New Scientist* magazine remarked, "Could the next battleground in the ID movement's war on science be the brain?"[29] We will get to that subject in chapter 8 and see why many believers are putting their stock in brain science.

DAWKINS'S DISPROOF

Perhaps the greatest misunderstanding, or misrepresentation, by theists of the arguments made by Dawkins in *The God Delusion* is with his scientific argument for God's nonexistence. I bring it up here because it relates to the claims of intelligent design.

In *God Is No Delusion: A Refutation of Richard Dawkins*, Thomas Crean asserts, "Professor Dawkins offers only one philosophical argument for atheism. It could be called 'the argument from complexity.' His idea is this: if a being existed with the attributes generally said to belong to God, such a being would be complex, and therefore require a cause."[30] Actually, this is not the only "philosophical argument" in Dawkins's book, which contains a discussion of most of the arguments given for God's existence and why they do not work. Furthermore, Crean gets Dawkins's point wrong. No one alive on this planet has written more than Richard Dawkins about how evolution by natural selection provides the mechanism by which complex biological systems develop from simpler systems. This is not just true for biology but for all physical systems composed of many particles, where a simple system such as water vapor can spontaneously change into a complex system such as a snowflake.

As we have just seen, this view has been challenged in recent years by the intelligent design movement, which claims that biological and other physical systems are too complex to be produced by simple, natural means and require an intelligent cause. What Dawkins has said is: All right, if that's the case, then the intelligent designer is even more complex and requires an even more intelligent cause. Of course Dawkins doesn't believe that for one minute and sees the universe as showing no evidence for design.

NOTES

1. My numerous papers on fine-tuning are linked at http://www.colorado.edu/philosophy/vstenger/anthro.html.

2. John D. Barrow and Frank J. Tipler, *The Anthropic Cosmological Principle* (Oxford: Oxford University Press, 1988).

3. Leonard Susskind, *Cosmic Landscape: String Theory and the Illusion of Intelligent Design* (New York: Little, Brown, 2005), p. 186.

4. Aurélian Barrau, "Physics in the Multiverse," *PhysicaPlus*, http://physicaplus.org.il/zope/home/en/1223032001/barrau_en/ (accessed December 1, 2008).

5. Rich Deem, "Evidence for the Fine Tuning of the Universe," God and Science, http://www.godandscience.org/apologetics/designun.html (accessed November 27, 2008).

6. Hugh Ross, *The Creator and the Cosmos: How the Greatest Scientific Discoveries of the Century Reveal God* (Colorado Springs, CO: NavPress, 1995).

7. Reasons to Believe, http://www.reasons.org (accessed December 20, 2008).

8. Evidence for Design, http://www.reasons.org/resources/apologetics/index.shtml#design_in_the_universe (accessed December 20, 2008).

9. The Heisenberg uncertainty principle states that the product of the uncertainty in the component of momentum along a given coordinate axis and the uncertainty in the position along that axis must be greater or equal to Planck's constant divided by 4π. The principle applies to other pairs of variables as well. The "uncertainty" is defined as the statistical standard deviation of measurements from the average value in an ensemble of measurements.

10. Peter Schneider, *Extragalactic Astronomy and Cosmology: An Introduction* (Berlin: Springer, 2006), p. 163.

11. In the actual study I also varied the strong nuclear force strength, but this does not have a direct role in the lifetime of stars.

12. Victor J. Stenger, *The Unconscious Quantum* (Amherst, NY: Prometheus Books, 1995), pp. 230–39; Victor J. Stenger, "Natural Explanations for the Anthropic Coincidences," *Philo* 3 (2000): 50–67. The simulation program, called "MonkeyGod," can be executed online at http://www.colorado.edu/philosophy/vstenger/Cosmo/monkey.html.

13. Anthony Aguire, "The Cold Big-Bang Cosmology as a Counter-Example to Several Anthropic Arguments," *Physical Review* D64 (2001): 083508; Roni Harnik, Graham D. Kribs, and Gilad Perez, "A Universe without Weak Interactions," *Physical Review*, available from arXiv.org, http://arxiv.org/abs/hep-ph/0604027 (accessed November 27, 2008); Fred C. Adams, "Stars in Other Universes: Stellar Structure with Different Fundamental Constants," to be published in *Journal of Cosmology and Particle Physics*, available from arXiv.org, http://arxiv.org/abs/0807.3697 (accessed November 27, 2008).

14. Richard Swinburne, "Argument from the Fine-Tuning of the Universe," in *Modern Cosmology & Philosophy*, ed. John Leslie (Amherst, NY: Prometheus Books, 1998), pp. 166–67; John Leslie, "The Anthropic Principle Today," in ibid., p. 291.

15. Martin J. Rees, *Just Six Numbers: The Deep Forces That Shape the Universe* (New York: Basic Books, 2000).

16. William Paley, *Natural Theology; Or, Evidences of the Existence and Attributes of the Deity*, 12th ed. (London: Printed for J. Faulder, 1809).

17. Ronald L. Numbers, "Darwinism, Creationism, and Intelligent Design," in *Scientists Confront Intelligent Design and Creationism*, ed. Andrew J. Petto and Laurie R. Godfrey (New York: Norton, 2007), p. 46.

18. John Paul II, "Message to Pontifical Academy of Sciences," Catholic Information Network, October 22, 1996, http://www.cin.org/jp2evolu.html (accessed December 20, 2008).

19. Ibid.

20. Michael J. Behe, *Darwin's Black Box: The Biochemical Challenge to Evolution* (New York: Free Press, 1996).

21. Mark Perakh, *Unintelligent Design* (Amherst, NY: Prometheus Books, 2004); Matt Young and Taner Edis, *Why Intelligent Design Fails: A Scientific Critique of the New Creationism* (New Brunswick, NJ: Rutgers University Press, 2004); Petto and Godfrey, eds., *Scientists Confront Intelligent Design and Creationism.*

22. William A. Dembski, *Intelligent Design: The Bridge between Science & Theology* (Downers Grove, IL: InterVarsity Press, 1999).

23. Victor J. Stenger, *Has Science Found God?* (Amherst, NY: Prometheus Books, 2003), pp. 102–10.

24. All trial documents are from case no. 04cv2688 in the United States District Court in the Middle District of Pennsylvania; they are available at http://en.wikipedia.org/wiki/Kitzmiller_v._Dover_Area_School_District_trial_documents (accessed November 29, 2008).

25. Bradley Monton, "Is Intelligent Design Science? Dissecting the Dover Decision," *PhilSci-Archive*, http://philsci-archive.pitt.edu/archive/00002592/ (accessed November 29, 2008).

26. The Arkansas case and Laudan's response are discussed in detail in *Has Science Found God?* pp. 53–62.

27. Barbara Forrest and Paul R. Gross, *Creationism's Trojan Horse: The Wedge of Intelligent Design* (Oxford: Oxford University Press, 2004).

28. The Discovery Institute's "wedge strategy" currently can be found at http://www.public.asu.edu/~jmlynch/idt/wedge.html (accessed November 29, 2008). This has been removed from the institute's Web site but its authenticity is documented. See James Still, "The Wedge Strategy Three Years Later," http://www.secweb.org/asset.asp?AssetID=200 (accessed November 29, 2008).

29. Amanda Geffer, "Creationists Declare War over the Brain," *New Scientist* 2679 (October 25, 2008): 46–47.

30. Thomas Crean, *God Is No Delusion: A Refutation of Richard Dawkins* (San Francisco: Ignatius Press, 2007) p. 10.

5. HOLY SMOKE

Men will never be free until the last king is strangled in the entrails of the last priest.

—Denis Diderot (d. 1784)

THE DARK BIBLE

Most philosophers of religion agree that the strongest case against the existence of God is the problem of evil. And, the strongest case against religion is its unbroken history as a major source of the most horrible evils that the world has seen.

In his 2004 book, *God against the Gods: The History of the War between Monotheism and Polytheism*, author Jonathan Kirsch notes, "Contrary to the conventional wisdom of Judaism, Christianity, and Islam, the world of classical paganism was *not* steeped in sin."[1] Religious liberty and diversity were core values of classical polytheism. When one nation defeated another in war, the losers were expected to worship the conquerors' gods but were still allowed to continue to worship their own. Polytheism did not have holy wars, inquisitions, and crusades. These were all the products of monotheism.

Kirsch regards the events of September 11, 2001, as a reminder of the three-thousand-year-old conflict between monotheism and polytheism:

> The men who hijacked and crashed four civilian airliners were inspired to sacrifice their own lives, and to take the lives of several thousand "infidels," because they had embraced the simple but terrifying logic that lies at the heart of monotheism: if there is only one god, if there is only one way to worship that god, then there is only one fitting punishment for failing to do so: death.[2]

Although the bloodiest acts of violence in the name of God nowadays seem to come from the Islamic world, the roots of religious terrorism are to be found not in Islam but in the early pages of the Bible.[3]

The story begins with the Old Testament, where the God YHWH is without mercy in his demands for the blood of those who do not worship him:

> If thy brother, the son of thy mother, or thy son, or thy daughter, or the wife of thy bosom, or thy friend who *is* as thine own soul, entice thee secretly, saying, Let us go and serve other gods, which thou hast not known, thou, nor thy fathers; *namely*, of the gods of the people which *are* round about you, nigh unto thee, or far off from thee, from *one* end of the earth even unto the *other* end of the earth, thou shall not consent to him nor hearken unto him; neither shall thine eye pity him, neither shalt thou spare, neither shalt thou conceal him: but thou shalt surely kill him; thine hand shall be first upon him to put him to death, and afterward the hand of all the people. And thou shall stone him with stones, that he die; because he hath sought to thrust thee away from the LORD thy God, which brought thee out of the land of Egypt, from the house of bondage. And all Israel shall hear, and fear, and shall do no more any such wickedness as this is among you. (Deut. 13:6–11)

The Jews would occasionally slip into the worship of other gods, notably the ever-popular Baal:

> And Israel joined himself unto Ba'al-pe'or: and the anger of the LORD was kindled against Israel. And the LORD said unto Moses, Take all the heads of the people and hang them up before the LORD against the sun, that the fierce anger of the LORD may be turned away from Israel. (Num. 25:3–4)

Of course, Israel's enemies were also given no mercy, not even the innocent:

> And it came to pass, that at midnight the LORD smote all the firstborn in the land of Egypt, from the firstborn of Pharaoh that sat on his throne unto the firstborn of the captive that *was* in the dungeon; and all the firstborn of cattle. And Pharaoh rose up in the night, he, and all his servants, and all the Egyptians; and there was a great cry in Egypt; for *there was* not a house where *there was* not one dead. (Exod. 12:29–30)

And the Lord showed no pity either for the people of Betshe'mesh:

> And he smote the men of Betshe'mesh, because they had looked into the ark of the LORD, even he smote of the people fifty thousand and threescore and ten men: and the people lamented, because the LORD had smitten *many* of the people with a great slaughter. (1 Sam. 6:19)

YHWH did not spare Israel his wrath when King David ignored a command:

> So the LORD sent pestilence upon Israel: and there fell of Israel seventy thousand men. (1 Chron. 21:14)

Under God's orders, Moses made war against the Midianites with an army of twelve thousand. They slew all the Midian kings and all the men, destroyed their cities, and took the women and children captive. When his captains brought the spoils and the women and children before Moses, he said:

> Have ye saved the women alive? Behold, these caused the children of Israel, through the counsel of Balaam, to commit trespass against the LORD in the matter of Pe'or, and there was a plague among the congregation of the LORD. Now therefore kill every male among the little ones, and kill every woman that hath known man by lying with him. But all the women children, that have not known a man by lying with him, keep alive for yourselves. (Num. 31:15–18)

The Bible sure speaks for itself. I have quoted only a tiny fraction of the passages that speak of terrible atrocities committed by the God that billions of people still claim to worship.[4] I need not repeat the story for the Qur'an, which renames YHWH as Allah, the most merciful, the most compassionate.[5]

Most of today's faithful ignore these passages, and many other similar passages. Christians rarely read the Old Testament. However, they cannot simply wash away this history. It provides evidence by which we can test the claim that a benevolent God is in charge of the world. You would expect such a world to show signs of goodness and moral perfection. Instead we find atrocities and gratuitous suffering, which seem to falsify the hypothesis of a morally perfect God.

But YHWH is still the Christian God. Somehow the cruel, vindictive God of the Old Testament reformed himself to become the personal, caring, loving God that is talked about from the pulpit in today's Protestant megachurches. More likely, God has always been an illusion, and the form of that illusion is what has changed as time has gone by.

WHAT DID JESUS DO?

I once heard a prominent Methodist minister in Colorado Springs debate the existence of God. His opponent, philosopher Keith Parsons, was so good that he almost convinced the minister that God did not exist. Well, maybe not. But the best the minister could say in response was that whether or not God existed, Jesus Christ was someone who was worthy to follow. Parsons even challenged that, referring to the fact that Jesus taught that only through him can one enter heaven.

In his book *Godless*, Dan Barker mentions several examples directly from the New Testament where Jesus hardly behaves like the perfectly good and loving role model he is always assumed to be by Christians.[6]

For example, Jesus encouraged the beating of slaves:

> And that servant [Greek *doulos* = slave], which knew his Lord's will and prepared not *himself*, neither according to his will, shall be beaten with many stripes. But he that knew not, and did not commit things worthy of stripes, shall be beaten with few *stripes*. (Luke 12:47–48, KJV)

Jesus never spoke out against poverty. Some of his disciples objected to the waste of a costly ointment used to anoint Jesus' head, "for it might have been sold for more than three hundred pence, and have been given to the poor" (Mark 14:5, KJV). Jesus replied, "For ye have the poor with you always, and whensoever ye will ye may do them some good: but me ye have not always" (Mark 14:7, KJV). Pretty selfish and arrogant for a benevolent God, wouldn't you say?

Jesus was not exactly nonviolent. In the Parable of the Ten Pounds, Jesus tells of a Lord whose people hated him. Barker, who has studied the Bible far more deeply than I have, says that Jesus is comparing the "Lord" in the parable to himself when he says: "But those mine enemies, which would not that I should reign over them, bring hither, and slay *them* before me" (Luke 19:27, KJV).

Jesus shows signs of paranoia when he says, "He that is not with me is against me" (Matt. 12:30, KJV). Finally, let me mention his description of hell, which is one of the few original ideas in the New Testament:

> The Son of Man shall send forth his angels, and they shall gather out of his kingdom all things that offend, and them which do iniquity; and shall cast them into a furnace of fire: there shall be wailing and gnashing of teeth. (Matt. 13:41–42, KJV)

Just a few decades ago Protestant preachers threatened hellfire and brimstone to their cowered congregations. Today's congregations do not cower or grovel. Hands waving in the air, the modern Pentecostal churchgoer sings herself into a self-induced trance similar to that achieved by Muslims, Catholics, and Jews with their highly ritualized prayers. The newcomers have discovered an ancient fact. The religious experience is personal, self-centered, and pleasurable. It doesn't matter what the Bible says, what theologians argue, or what history tells us. That can be ignored.

Well, the new atheists are going to keep the feet of moderate as well as extremist theists to the fire about the evils of institutionalized religion until they are willing to denounce them. It is our view that their own irrational beliefs feed the extremism of others.

HISTORICAL HORRORS

In his 2002 book journalist James A. Haught (not to be confused with theologian John F. Haught, whom I have quoted often) has compiled a list of atrocities committed in the name of God.[7] Haught lists forty-three such atrocities.[8] Let me first mention a few of his examples to provide an overview, then we will discuss the points where theists take issue in more detail.

The First Crusades began in 1095 with the hacking to death or burning alive of thousands of Jews in the Rhine Valley in Germany as the crusaders gathered there. The religious legions proceeded to plunder their way two thousand miles to Jerusalem, killing virtually every inhabitant of the towns along the way.

In launching the Second Crusade, Saint Bernard of Clairvaux declared: "The Christian glories in the death of a pagan, because thereby Christ himself is glorified."

In 1191, during the Third Crusade, Richard the Lion-Hearted captured Acre and ordered the slaughter of three thousand men, women, and children while a bishop intoned blessings.

In 1209, Pope Innocent III ordered a crusade against the Albigenses in southern France, a group of Christian heretics especially noted for their asceticism and piety. When the city of Beziers fell, soldiers could not distinguish between the heretics and the good Catholics. So the papal adviser commanded: "Kill them all, God will know his own."

The Albigensian heresy led Pope Innocent IV to establish the Inquisition. Dominican priests devised fiendish machines of torture that were used on thousands in France and Spain.

In the 1400s the Inquisition changed its focus to witchcraft and thousands of women were tortured into confessing and then burned or hanged. Witch hysteria raged for three centuries with estimates of the number executed ranging from a hundred thousand to two million.

The Thirty Years' War between Protestants and Catholics began in 1618 and turned central Europe into a wasteland. Germany's population dropped from eighteen million to four million, according to one estimate. Another estimate is that 30 percent of the population, 50 percent of the men, were killed.

And this wasn't the end of Protestant-Catholic violence. Three thousand lives have been lost in Ulster since that age-old hostility turned violent again in 1969.

Of course, Christians were not the only ones killing for religious reasons. Central and South American cultures that flourished between the eleventh and sixteenth centuries engaged in human sacrifice. The Aztecs killed twenty thousand people yearly to appease the gods. The Incas sacrificed two hundred children in a single ceremony. In Tibet, priests performed ritual killings. The Dravidians in India sacrificed a male child to Kali every Friday evening.

Islamic jihads killed millions over twelve centuries. In the early 1900s Muslim Turks waged genocide against Christian Armenians and Greeks. Muslim countries today still practice barbaric punishments such as amputations for thievery and stoning to death for adultery.

Even Mahatma Gandhi was unable to stop the killing between Muslims and Hindus that took an estimated million lives when India became independent from Britain in 1947. Gandhi himself was assassinated by a Hindu fanatic who thought he was too pro-Muslim.

Imagine no religion. Notice that religion is the only feature that distinguishes most of these groups. Why would Catholic and Protestant Europeans kill each other if there were no religion? Why would Catholic and Protestant Irish kill each other if there were no religion? Why would Hindu and Muslim Indians kill each other if there were no religion? Why would Jewish and Muslim Semites kill each other if there were no religion?

ATHEIST HORRORS?

While most theists accept that religion has resulted in much unnecessary suffering in history, they argue that atheists, notably Stalin, Mao, and Hitler, killed more people in the twentieth century alone than were killed for religious reasons in all the previous centuries put together.

In his 2008 book, *The Irrational Atheist*, libertarian writer Vox Day claims that the various European inquisitions resulted in the deaths of no more than 6,405 people (exactly), compared to the millions killed by atheist regimes in the last century.[9] He dismisses the Crusades as not being a holy war at all but "an object lesson in the tragedy that can take place when religious ideals are perverted by Man's sinful nature."[10]

In their books, the new atheists have demonstrated that Hitler was not an atheist and provided evidence of the complicity of the Catholic Church with Nazism. Day argues that Hitler was not a Christian, despite the fact that he was a baptized Catholic and was never excommunicated by the Church. The facts are these. Hitler was not about to allow any rival institution have any significant power in Germany and was drawing up plans for a National Reich Church, whose doctrines would differ significantly from those of the Catholic Church. Day concludes, "Hitler was neither a Christian or an atheist."[11] Not being an atheist, the six million Jews he killed in the Holocaust, plus the millions of other deaths for which he was responsible, can hardly be assigned to the atheist side of the balance sheet.

Day and other critics have not successfully countered another point raised by the new atheists with respect to the Holocaust.[12] Again, imagine no religion. In that case, the Jews would never have developed as a separate people, forever stigmatized as the killers of Jesus. They would have melded in with the rest of society like the Canaanites from which they originally separated. In 1936 Hitler asked Bishop Berning of Osnabrüch, Had the Church not looked at Jews as parasites and shut them up in ghettos? "I am only doing what the church has done for fifteen hundred years, only more effectively."[13]

So chalk up at least six million twentieth-century deaths to religion.

Now, what about Stalin and other communist dictators? Day and other anti-atheists refuse to accept the new atheist argument that the communists did not commit their murders in the name of atheism. As Dawkins put it, "What matters is not whether Hitler and Stalin were atheists, but whether atheism systematically influences people to do bad things. There is not the smallest evidence that it does."[14] Day attacks Dawkins head-on, not neglecting the ad hominems: "Again Dawkins reveals his historical ignorance, and again he demonstrates that he is not so much a bad scientist as an atheist propagandist who has abandoned science altogether. For there is not only the smallest evidence that atheism correlates with people doing bad things, the evidence is so strong that it is almost surely causal."[15]

Having said that, Day knows he needs to come up with some powerful data. So he refers to the 1,781 Christian kings who ruled from 392 CE, the beginning of Christendom, to 800 CE, when Charlemagne was crowned Holy Roman Emperor. "Although those 1,781 Christian rulers, like rulers everywhere, engaged in wars and indulged in murders and committed plenty of other deplorable deeds, very, very few of them ever engaged in a systematic act of mass murder that can be reasonably described as anything approaching the crimes of the sort committed by Stalin."[16] And that is Day's "evidence" that is "so strong that it is surely causal."

Instead of looking at the kings of Christendom (why stop with Charlemagne?), whose motives would more likely be their own greed and thirst for power than religion, why not consider the popes who were the true rulers of Europe in the centuries known as the Dark Ages (curiously coincident with Christendom) and whose motives can surely be attributed, partly or wholly, to religion.

For starters, an estimated two hundred thousand to one million people were systematically massacred during the Albigensian Crusade (1209–1229), ordered by Pope Innocent III. Stalin killed perhaps ten million from a Soviet population many times greater than that of southern France in 1209,[17] so Innocent III was a far greater criminal than Stalin if you consider the percentages.

The crimes committed by the descendants of Peter, the "rock" upon which Jesus supposedly built his Church, was chronicled by a former priest, Peter De Rosa, in his 1988 book, *Vicars of Christ: The Dark Side of the Papacy*.[18] I will just mention two more not-so-saintly popes in the same mold as Innocent III.

In spring of 1299, Pope Boniface VIII flattened the city of Palestrina, praised by Horace for its beauty and containing one of the palaces of Julius Caesar, killing a reported six thousand people. Boniface accused the town of being "disloyal." For this act, Dante buried Boniface VIII in the Eighth Circle of Hell, head down in fissures in the rock.[19]

In 1377, the future (disputed) Pope Clement VII was papal legate to the town of Cesena on the Adriatic. Locals objected to his mercenaries raping their women and had killed some of the guilty men. After persuading the townspeople to lay down their arms, he sent in a mixed English and Breton force to slaughter all eighteen thousand inhabitants, including children.[20]

STALIN AND ATHEISM

In *The Plot to Kill God*, sociologist Paul Froese claims that the Soviet Union waged a relentless war on religion that he attributed to the "violence of atheism" and, despite this effort, it failed to eradicate faith. It turns out there is another side to the story, supported by hard evidence that disputes Froese's conclusion. In fact, since 1943 the Soviet Union has supported the Russian Orthodox Church. Despite that support, only 25 percent of Russians today are believers.

In his 2005 book, *Fighting Words: The Origins of Religious Violence*, Iowa State Religious Studies professor Hector Avalos reports on his examination of archival materials released after the fall of the Soviet Union. He found no evidence that Stalin killed because of atheism. Rather, the data indicate that Stalin's genocide was driven by the politics of forced collectivization.[21]

Historian Edvard Radzinsky has also examined the archives and found several examples where Stalin supported the Church.[22] Priests were allowed in the Soviet camps during the sieges of Stalingrad and Leningrad. The icon Our Lady of Kazan was carried in a procession on the streets of Leningrad. On September 8, 1943, with Stalin's permission, Metropolitan Sergius was elected patriarch of the Russian Orthodox Church (ROC), thus normalizing Soviet relations with the Church.

Avalos writes that in 1944 about 144,000 people attended Easter services in Moscow churches. A single church in the city of Kuibyshev recorded 22,045 baptisms in 1945 and 5,412 in the first three months of 1946. Stalin provided financial support to the ROC of 550,000 rubles in 1946 and 3,150,000 rubles in 1947.[23]

As we will see in the final chapter, governments have always used religion to help them maintain power and keep the people in line. Although officially atheist because of its commitment to Marxist dogma, the Soviet Union under Stalin and his successors, up to and including the current Russian regime, have recognized the benefit the Russian Orthodox Church had always provided the czars and put it to use to help maintain their own power.

FOLLOWING ORDERS

Perhaps some insight into how killing in the name of God comes so easy to true believers can be gained by looking at the newest of the large religions in the world, the Church of Jesus Christ of Latter-Day Saints (LDS), known as the Mormons. The noted author Jon Krakauer has written brilliantly about the church in his 2003 best seller, *Under the Banner of Heaven*.[24] His story builds around a vicious murder carried out, according to the perpetrators, upon God's order. In what follows I will give a lengthy summary of Krakauer's book, which hardly does it justice, with some additional information from other sources.[25]

On July 24, 1984, in American Fork, Utah, Brenda Lafferty and her infant daughter Erica were murdered by her brother-in-law, Dan, assisted by his brother Ron and two accomplices. The killers were all members of a polygamist sect that was an outlawed outgrowth of the LDS. In preparing for his book, Jon Krakauer held prison interviews with Dan Lafferty and other Mormons. Dan claimed the killings were ordered by God as revealed to Ron.

Mormonism is a uniquely American religion and, thanks to a vigorous missionary system, is one of the fastest-growing religions in the world with about thirteen million adherents—about the same number as Jews. By studying the Mormons, whose origins lie so much closer to us in time than any other large religion, we have the ability to gain considerable insight into how religions get started as well as to better comprehend their almost universal tendency to be murderous.

THE SAGA OF JOSEPH SMITH

The LDS Church was founded by Joseph Smith Jr., who was born in 1805. As a young man, Smith lived in upstate New York in the Palmyra area near Rochester. He began his career as a charlatan and treasure hunter. He and his father, Joseph Smith Sr., practiced a form of dowsing in which they used a divining rod to search for hidden Indian treasure, Spanish gold, and silver mines that were rumored to be in the area. Joseph Jr. also used a "seer stone," which was something like a crystal ball. In 1826, at age twenty-one, he was convicted of being "a disorderly person and an imposter."

The Smiths were invariably unsuccessful in their diggings until Joseph Jr. discovered the Book of Mormon. Smith claimed that on September 21, 1823, when he was seventeen, an angel named Moroni appeared to him and revealed the location of gold plates on the hill Cumorah near his home. His translation of the plates became the Book of Mormon, which he published in 1830. Unfortunately Smith did not keep the plates but said he returned them to Moroni.

Smith claimed the plates were in the ground since 421 CE and were written in "reformed Egyptian" characters. They told the story that the American Indians were descended from remnants of the lost tribes of Israel who had sailed to the New World in about 600 BCE, just before the last Babylonian conquest. They were headed by a virtuous man name Lehi. After Lehi's death the tribe separated into two clans led by his sons Nephi and Laman. The Nephites were pious and industrious while the Lamanites were idle and full of mischief. The Lamanites so annoyed God that he turned their skin "dark and loathsome" in order to punish them.

In 34 CE, after his resurrection and ascension, Christ visited America, appeared to both clans, and shared a new gospel urging them to live together peaceably. This they did for several hundred years until the Lamanites began to backslide into unbelief and idolatry.

Eventually hostilities broke out into a full-blown war, culminating in 400 CE when the Lamanites slaughtered all 230,000 Nephites. The leader of the Nephites in these final battles was a heroic and wise person named Mormon, whose son Moroni was the last surviving Nephite. It was Moroni, now an angel, who appeared to Joseph Smith in 1823 with the account of this history.

Starting with fifty members in August 1830, Smith formed the Mormon Church, which grew rapidly under its charismatic prophet. Of course, he claimed the authority of divine revelation, as has the founder of just about every other religion the world has seen with the exception of Buddha. But Smith came up with a unique teaching that made the new religion exceptionally attractive: Every "Saint" had the same privilege of communicating directly with God. Of course, Smith himself kept up a steady stream of revelations defining Church doctrine that generally took precedence over the revelations of the local barber. The murders of Brenda and Erica Lafferty were not the first example of violence within Mormonism brought about by their belief in individual revelations.

Realizing that he would lose control of his followers if everyone were a revelator, within a year after founding the LDS Church Smith had another revelation from God saying that "No one shall be appointed to receive commandments and revelations in this church excepting my servant Joseph Smith Jr." But, as Krakauer says, the "genie was already out of the bottle.... People liked talking to God directly, one-on-one, without intermediaries."[26] The result has been a constant splitting off of schismatic Mormon sects, about two hundred total since 1830.

In December 1830 Smith moved his clan of followers from New York to Kirkland, Ohio, which was a way station to the place where the Lord revealed he had consecrated for the Saints—Jackson County, Missouri. Followers poured into Jackson County, much to the distress of its existing inhabitants. Armed mobs terrorized the Saints, whose population in Missouri by 1838 had reached ten thousand. With the encouragement of the Missouri legislature and the federal government, the Mormons moved north to the more sparsely populated Caldwell County and built a town they called Far West.

But trouble still erupted in adjacent Davies County. For the first time Mormons began to fight back, which only branded them as enemies to be exterminated for the sake of public peace. On October 30, 1838, the Missouri militia launched a surprise attack on a Mormon settlement, killing eighteen Saints. The Mormons were forced to pay compensation and to leave Missouri altogether. They settled just across the Mississippi in a swampy town in Illinois they renamed Nauvoo. Hundreds died of malaria and cholera, including the prophet's father and one brother, until the swamps were drained. Still, in five years the industrious Mormons had created a major city. In December 1840 the Illinois General Assembly passed a charter that set up Joseph as de facto ruler of his own city-state.[27] He commanded a militia that had almost half the number of men as the whole US Army. He even ran for president of the United States in 1844.

Smith had a taste for women, and one of his most important revelations was that God wanted men to have multiple wives. After some delay, he formally codified this doctrine on July 12, 1843. Eventually Smith accumulated forty or so wives, while also regularly visiting Nauvoo's houses of ill repute. He was often denounced for his behavior but managed for a while to keep it under control (the criticism, not his behavior).

Influential Saints of high moral character urged him to renounce polygamy, and surrounding non-Mormons were agitated by the practice. In 1844 violence broke out that resulted in Smith's being arrested by state authorities for destroying the office of the local newspaper that had editorialized against him. Joseph and his brother Hyrum were incarcerated in a jail in Carthage, Illinois. On the afternoon of July 27, 1844, 125 militiamen from the anti-Mormon town of Warsaw attacked the jail and Joseph and Hyrum were shot to death. Wounded seriously was Apostle Joseph Taylor, who would become the Church's third president, succeeding Brigham Young in 1877. Nine years later Taylor received a revelation affirming the righteousness of plural marriage.

THE MOUNTAIN MEADOWS MASSACRE

Brigham Young took over the leadership of the LDS and moved the faithful to the territory of Utah. He ruled with an iron hand and officially sanctioned polygamy in 1852. He reportedly had dozens of wives, sixteen of which gave birth to his fifty-seven children. Needless to say, he has many descendants, including pro-football hall-of-famer Steve Young.

Although appointed governor of the territory, Brigham Young ruled a theocracy that ignored federal laws such as those against polygamy in the territories. His militia, the Nauvoo Legion, harassed federal agents working in Utah. When President James Buchanan installed a new governor for the territory, Young declared martial law and a federal army headed for Utah to wrest control of the territory from Young.

Krakauer tells the story of the Mountain Meadows massacre, which took place in Washington County at this time, in 1857. Unaware of the tense situation in Utah, an unusually large and wealthy wagon train from Arkansas with 140 individuals, nine hundred head of cattle, and a prize racehorse worth many hundreds of thousands of today's dollars crossed into Utah on its way to California. They were also rumored to be carrying a strongbox filled with thousands of dollars in coins. It was known as the Francher Party. They entered into desperate country. A plague of crickets and drought had put many Saints on the edge of starvation. Their religion taught that stealing from the godless was righteous, so the wagon train looked ripe for picking. Furthermore, an important Mormon apostle, Parley Pratt, had just been savagely killed in Arkansas, near where the Francher train originated, and the Mormons were seeking revenge.

As the wagon train entered Utah, Apostle George A. Smith, first cousin to Joseph and general in the Nauvoo Legion, held a powwow with hundreds of Paiute Indians about twenty miles from Mountain Meadows. He told the Indians that the Americans had a large army just to the east of the mountains and intended to kill all of the Mormons and Indians in the Utah territory. Smith urged the Indians to get ready for war against all of the Americans and to "obey what the Mormons told them to do—that this was the will of the Great Spirit."[28]

The Indian agent and interpreter at the powwow, Mormon John D. Lee, said twenty years later that he always believed General Smith had been sent there to incite the Indians to exterminate the wagon train and "was sent from that purpose by the direct command of Brigham Young."[29]

After a five-day battle with Indians and, it is believed, Mormon militiamen dressed as Indians, Lee approached the wagon train with a proposal that they give up their weapons in return for safe passage. Having little other choice, they agreed and the women and children were sent ahead. The men of the party were led away by militiamen and each one was shot or clubbed to death. The militiamen guarding the women and children, disguised as Indians, killed all but seventeen children under age five—too young to be able to witness against the Saints. These were taken in by Mormon families. All told, about fifty men, twenty women, and fifty children or adolescents were killed.

After a minimal burial in shallow graves that scavengers easily uncovered in days, the Saints gathered round and offered "thanks to God for delivering our enemies into our hands."[30]

After twenty years of cover-ups in which attempts were made to pin the blame on the Indians, John D. Lee, who had become the wealthiest man in southern Utah with several homes and eighteen wives, was eventually hunted down and placed on trial in Beaver, Utah. No other participant was ever brought to justice. In what Krakauer likens to the O. J. Simpson verdict, the jury deadlocked and Lee was not convicted.

This caused a national outcry that led Brigham Young to cynically stop blaming the Indians and put the full onus on Lee as a scapegoat. At a second trial in 1876, Young screened the jurors and Lee was found guilty of first-degree murder. Lee spent his last days awaiting execution writing his life story, *Mormonism Unveiled*, which posthumously became a national best seller. Lee was executed by firing squad on March 23, 1877.

THE FUNDAMENTALISTS

Utah was admitted as the forty-fifth state in the Union on January 4, 1896, on condition that it outlaw polygamy. While the largest Mormon sect, the LDS Church, grudgingly stopped the practice, which was considered by most members to be a major tenet of Mormonism, other Mormon sects continued the practice. Even to this day, attempts to enforce the law against polygamy have been unsuccessful because of lack of popular support.

Even though the majority of LDS members today are dutifully monogamist, many came from polygamist families and are not ready to renounce their ancestors nor what they believe was a legitimate revelation from God to Joseph Smith.

The largest sect of Mormon polygamists is the Fundamentalist Church of Jesus Christ of Latter-Day Saints, centered in Colorado City, Arizona, near the Utah border. Its members, and those of other polygamist sects, believe LDS sold out on one of the most sacred covenants of their religion.

But any objective observer can see that polygamy as practiced by Mormons today is nothing less than female slavery. Here's how one woman described her experience growing up in Colorado City to a *Denver Post* reporter:

> As a little girl, Laura Chapman taped the words "Keep Sweet" on her bathroom mirror to remind herself how to get by in this remote, polygamous community. Keeping sweet, Chapman says, meant staying silent as her father molested her starting at age 3. It meant hiding her secret from her 30 brothers and sisters. It meant being lashed with a yardstick by one of her father's four wives. It meant having to quit school at age 11, then work without pay in a store owned by her church's prophet.
>
> Keeping sweet meant being forced into marriage at age 18 to a man she didn't know, let alone love. It meant having a baby every year. It meant walking 10 paces behind her husband. And, above all, it meant smiling, sweetly through her pain.
>
> "We were just little girls in odd clothes and funny hair who thought we were going to hell if we didn't obey," recalled Chapman, now 38, who has made a new life in Longmont [Colorado] since fleeing 10 years ago with her five children. "Who would think, right here in the United States of America fathers are trading their daughters away like trophies? It's brainwashing and slavery. It's a complete system of organized crime right in our backyard that for some reason the government has simply chosen to ignore."[31]

Abandoning the practice of polygamy has helped the primary LDS Church to become regarded by most Americans as a mainstream Christian faith, despite its highly unconventional, indeed bizarre, theological foundation. But then, come to think of it, all faiths are bizarre.

Dan Lafferty was raised as a conventional Mormon. He led a rather typical life, training as a chiropractor (pseudoscience is quite common among Mormons) in California before moving back to Utah with his wife, Matilda. He started to look into the history of Mormon polygamy and came across a treatise called *The Peace Maker*, written by Udney Hay Jacob and printed by none other than Joseph Smith. *The Peace Maker* provided an elaborate rationale for polygamy, and Dan became convinced that the Lord had given him the knowledge that it was written by Joseph Smith and was a valid revelation. Most scholars today agree that Jacob was the author.

Dan quickly converted his household of Matilda and six children, including two from Matilda's previous marriage, to fundamentalism of the variety we saw in Colorado City. Here's how Krakauer describes their new life:

> Matilda was no longer allowed to drive, handle money, or talk to anyone outside the family when Dan wasn't present, and she had to wear a dress at all times. The children were pulled out of school and forbidden to play with their friends. Dan decreed that the family was to receive no outside medical care; he began treating them himself by means of prayer, fasting, and herbal remedies.[32]

Matilda testified her life had become "hellish." Dan proposed that he take Matilda's oldest daughter, his stepdaughter, as his wife. Instead, Dan married another woman.

In his fundamentalist zeal, Dan began to take various documents such as the Book of Mormon and *The Peace Maker* literally as overriding commands from God. This meant, in his mind, he had the right to ignore certain laws such as paying taxes and getting a driver's license, which got him in trouble with the law. He influenced his four younger brothers to join in his crusade, which became more overtly religious. Their wives complained about their increasingly miserable lives to Diana, the wife of the eldest Lafferty brother, Ron. One evening Ron, a dutiful Saint, sat down to reason with his brothers and in a few hours Dan had converted him to a fire-breathing fundamentalist. Diana soon went from being treated like a queen to a slave. All the wives dutifully submitted to the humiliation called for in *The Peace Maker* except one, Brenda, the wife of Allen, the youngest of the six brothers.

Brenda was intelligent, articulate, and assertive, and refused to go along. She was more sophisticated than the average fundamentalist wife, having majored in communications at Brigham Young University, and while there she anchored a television newsmagazine program on KBYU, the local Public Broadcasting Service affiliate.

In 1984 the brothers joined the new School of the Prophets that had been set up by a self-proclaimed prophet, Robert C. Crossfield, who called himself Onias. Onias had published a set of pamphlets called *The Book of Onias* containing his revelations from God that condemned modern Mormon leaders for defying the most sacred doctrines God had revealed originally to Joseph Smith. These included their enforcing the criminalization of plural marriage and their admission of black men into the Mormon priesthood.

Onias told them that he had received a revelation that God had singled them out as elect people. Ron was appointed bishop of the school's Provo chapter. By this time Diana had divorced him and moved with their six children to Florida, as far away as she could get. Ron had no income and was living out of his 1974 Impala station wagon.

On February 24, 1884, Ron became the first of Onias's students to receive a direct commandment from God. Others followed. All told, Ron received approximately twenty revelations that month and the next.

RON'S REVELATION

In a revelation received by Ron late in March, God commanded his servants and prophets to remove four individuals so his work could go forward. The first two were his brother's wife, Brenda, and her baby. The matter was to be taken care of as soon as possible.[33] In yet another revelation God implied that Dan was to do the actual killing.

In May 1984 Ron and Dan made an extensive motor trip visiting various fundamentalist communities in the American West and Canada. They returned in late July. On the morning of July 24 Dan got up, prayed, and prepared his weapons. The following is Dan's account of the story. Ron's differs somewhat, but Krakauer finds Dan's more credible. Dan, Ron, and two companions loaded the Impala and climbed in. After stopping for another weapon, they pulled into Allen and Brenda's driveway at about

1:30 in the afternoon. At first Ron had no answer to his knocks and Dan began to drive away, thanking God. But then Dan had further thoughts about his mission and turned and went back, this time knocking on the door himself. Brenda answered and Dan pushed his way in. The two scuffled, Brenda begging him not to hurt her baby. Dan was beating her with his fists as Ron came in the house. Brenda attempted to get away but Dan grabbed her and she fainted. Dan tied a vacuum cleaner cord tightly around her neck.

Dan then picked up a knife and found the baby's room. She was standing in the corner of her crib. As Dan described it from his cell:

> I spoke to her for a minute. I told her, "I'm not sure what this is all about, but apparently it's God's will that you leave this world; perhaps we can talk about it later." And then I set my hand on her head, put the knife under her chin like this, and I just...[34]

He had virtually decapitated her.

After wiping the knife clean, Dan went downstairs and slit Brenda's throat. He washed the knife a second time and told Ron, "Okay, we can leave now."

They still had two people to kill under God's orders. One was not home and they could not find the other's house, so the killers kept driving with the plan to go to Reno. Along the way, Dan and Ron's companions deserted them and the two brothers were apprehended in Reno on August 7.

Dan and Ron were tried separately, with Dan receiving two sentences of life imprisonment on January 15, 1985, after his jury was not unanimous on the death penalty.

Ron's trial began in April 1985. He refused an insanity defense and was quickly found guilty of two counts of first-degree murder. In 1991 the conviction was overturned by the Tenth US Circuit Court of Appeals in Denver, Colorado, which ruled that his paranoid delusions prevented him from interpreting the consequences of his conviction in a realistic way. This led theologians and others to mull over the potential consequences. It implied a secular version of sanity by suggesting that anyone who talks to God is crazy. Since the whole Mormon faith is based on talking to God, were they all legally insane?

Before his retrial, Ron underwent months of psychotherapy, and in

February 1994 it was ruled he was competent to stand trial. That trial did not begin until early 1996 and most of it was occupied with the defense trying to prove that Ron was deranged and the prosecution trying to show that Ron's beliefs, while unusually stated, were not really much different from those of billions of people in the world. The main prosecution witness was Noel Gardner, MD, a psychiatrist affiliated with the University of Utah Medical School. Here is how he explained the distinction between preposterous religious tenets and clinical delusion:

> A false belief isn't necessarily a basis for mental illness. [Most of humankind subscribes to] ideas that are not particularly rational.... For example, many of us believe in something referred to as trans-substantiation. That is when the priest performs the Mass, that the bread and wine become the actual blood and body of Christ. From a scientific standpoint, that is a very strange, irrational, absurd idea. But we accept that on the basis of faith, those of us who believe that. And because it has become so familiar and common to us, that we don't even notice, in a sense, it has an irrational quality to it. Or the idea of the virgin birth, which from a medical standpoint is highly irrational, but it is an article of faith from a religious standpoint.[35]

Gardner testified that Ron did not possess the symptoms of a schizophrenic but rather someone who exhibited the symptoms of *narcissistic personality disorder.*[36]

On April 10, 1996, the jury convicted Ron of first-degree murder. He was sentenced to death by either lethal injection or four bullets through the heart at close range. Ron chose the latter.[37] He is still on death row.

In the meantime, during his twenty-three years in jail, Dan has rethought his religious position. He has convinced himself that he is the "new Elijah," designated by God to recognize Christ when he kicks off the thousand-year reign of the kingdom of God.

LESSONS FROM MORMONISM

I have reviewed Krakauer's *Under the Banner of Heaven* in some detail because it is a beautifully written, fascinating, and deeply illuminating tale about a major religion that was created in America only 180 years ago. We can be so much surer of the history of the Church of Jesus Christ of Latter-Day Saints than other major religions, all of which go back over a thousand years to the point where attempts to place them on a secure historical footing is ludicrous by comparison. We know Joseph Smith existed. We cannot be anywhere near as sure for Muhammad and Jesus, for whom there is really little if any historical evidence, and we can almost say for sure that Abraham and Moses are not historical figures.

Here's what we know about Joseph Smith, the founder of Mormonism. We know he was a con man who took money from farmers on the dishonest promise of finding gold on their property, which he never did. He then hit on an even greater scam, claiming to find gold tablets that had been buried for centuries. The tablets told a fantastic story of America being populated by two tribes of Jews who sailed to the New World in 600 BCE. Of course, anyone with the slightest knowledge of history or archaeology knows that this story has absolutely no basis in fact. It is a good example of absence of evidence being evidence for absence. The lack of physical evidence for these events having taken place when such evidence should be present, along with many of the other dubious teachings of Mormonism, completely falsifies a religion practiced by thirteen million people worldwide. Mormonism is not just wrong, it is provably wrong.

The fact that today's Mormons, at least the great majority associated with the main LDS Church in Salt Lake City, are law-abiding, family-oriented, industrious members of society has made most Americans accept them as good Christians—on the edge of the mainstream but still worthy Americans. We have seen that they have had a violent history, but certainly all that violence was not caused by Mormons. The murders committed by fundamentalists described by Krakauer are hardly representative of the people who call themselves Mormons, but they do show us how the irrational thinking processes built into this religion, as well as other religions, have led to so much mayhem in history.

While I realize I am saying the obvious, which has been said many times by others, we have with the case study of the murders of Brenda Lafferty and her baby a detailed demonstration of how anyone who thinks he possesses absolute truth is capable of any horrible act, which is justified in his mind by that belief. Note that this not only explains the holy horrors of history, but also the terrible acts committed by Hitler, Stalin, and Mao. Hitler killed six million Jews for at least partially religious reasons. They killed

God. In any case, if there had been no religion there would not have been a group so separated from the rest of society that they stood out as different.

Stalin and Mao did not kill in the name of religion, but they killed in the name of another type of belief in absolute truth. This was not the truth of atheism, which no atheist proclaims as dogma to kill by, but the absolute truth that they had come to accept on the basis of the evidence of exploitation of workers by the capitalist societies, with at least a dollop of faith that their solution, the doctrine of communism, was the only answer. As Sam Harris, the first of the new atheists pointed out, it is faith that is a major source of evil in the world.

HOLY TERROR

I have noted that the New Atheism movement was most likely triggered by the events of September 11, 2001. Sam Harris admits this was his motivation for writing *The End of Faith.* Its market success, and that of the other best sellers by new atheists, was probably also a product of that horrendous day. Although President Bush and other leaders tried to gloss over the religious significance, insisting Islam was "a religion of peace," most Americans were struck by the sheer religious nature of the attacks. Only the most muddle-headed academics, such as Noam Chomsky (see chapter 1), blamed the violence on American oppression of Muslim nations.

Still, many critics accused the new atheists, and Harris in particular, of not having a sufficient understanding of Islam to draw the strong conclusion that Islam is an inherently violent religion whose members are driven to fanaticism by their faith. One such critic was my personal friend and colleague, physicist Taner Edis of Truman State University. Edis grew up in a secular family in Turkey and has studied Islam extensively. His book *An Illusion of Harmony* is an invaluable reference on science and Islam.[38]

In his commentary on Harris in *Free Inquiry* magazine, Edis objects to Islam being portrayed as a violence-obsessed religion based on quoting "verses of the Qur'an that promise sadistic punishments for unbelievers in the afterlife, urge fighting against infidels, and otherwise show an unhealthy preoccupation with vengeance and violence."[39] Edis explains that "Ordinary Muslims depend heavily on their local religious scholars, Sufi orders and similar brotherhoods, officially sanctioned clergy, and other mediating institutions. They hold the Qur'an sacred, but their understanding of what Islam demands comes through their local religious culture."

Whether murderous Muslims are influenced by their own reading of the Qur'an or by religious leaders, they still commit their murders in the name of Allah. Neither Harris nor any of the other new atheists condemns the great majority of Muslims as terrorists. But we hold them responsible nonetheless for the undeniable fact that their religion played an important role in the terror of September 11 and in the continuing warfare against modernity waged by Muslim extremists worldwide. The fact that the majority of Muslims do not read the Qur'an but learn their religion from their mullahs just demonstrates a fact about all religions, including Christianity. Christians do not read the Bible, either. If they did, they wouldn't be Christians. They listen to the selected verses read from the pulpit and taught in so-called "Bible study" sessions.

Bruce Lincoln, a professor of Divinity at the University of Chicago, has written about the religious implications of September 11 in *Holy Terrors: Thinking about Religion after September 11*.[40] He concludes: "It was religion that persuaded Mohamed Atta and eighteen others that the carnage perpetrated was not just an ethical act, but a sacred duty."[41]

Lincoln offers convincing evidence that the attacks were deeply rooted in Islamic thinking. That evidence is provided in appendix A of *Holy Terrors*, which presents the final instructions Mohamed Atta gave to the other hijackers, three copies of which have survived.[42]

No one reading these can possibly view Atta, at least, as a "freedom fighter" seeking to right injustices perpetrated on his people by the United States. The hijackers are urged to think of the Prophet, pray continuously, and read select suras from the Qur'an on what God has promised martyrs. They are to shower, shave excess hair, and wear cologne so they will be clean when they enter heaven.

They are to purify their souls from all unclean things and completely forget something called "this world" or "this life." Upon entering the aircraft they are to make a prayer and supplications. Atta tells them, "Remember that this is a battle for the sake of God."[43] And, as the planes near their targets:

> When the hour of reality approaches, the zero hour, [unclear] and wholeheartedly welcome death for the sake of God. Always be remem-

> bering God. Either end your life while praying, seconds before the target, or make your last words: "There is no God but God. Muhammad is His messenger. Afterwards, we will all meet in the highest heaven, God willing."[44]

In 2000, before the acts of Atta and his men, Mark Juergensmeyer wrote and published a highly acclaimed study of religious violence called *Terror in the Mind of God*.[45] In this he covered earlier perpetrators of terrorism including the Christians Mike Bray, Eric Robert Rudolf, and Timothy McVeigh; the Jews Yoel Lerner and Baruch Goldstein; the Muslims Mahmud Abouhalima and Abdul Aziz Rantisi; the Sikh Simranjit Singh Mann; and the Buddhist Nakamura.

Let me briefly summarize two terrorist acts Juergensmeyer studied in-depth and how the perpetrators rationalized them in terms of their personal beliefs.

In 1985, the Rev. Michael Bray of Bowie, Maryland, was convicted of destroying seven abortion facilities in Delaware, Maryland, Virginia, and the District of Columbia. He served prison time until 1989. Bray supported the 1994 killing by the Rev. Paul Hill of abortion provider Dr. John Britton and his escort. Bray justified both his and Hill's actions as defensive rather than punitive acts against the killing of babies. Hill found support in Psalms 91: "You will not be afraid of the terror by night, or of the arrow that flies by day."

Both Bray and Hill were associated with an extreme right wing of Christianity called Reconstruction Theology, whose proponents wish to create a theocratic state in preparation for Christ's return.[46]

The ancient town of Hebron lies on the West Bank and is occupied almost exclusively by Palestinian Arabs. The small population of Jews who live there are fervently religious and believe the land, with many sites they regard as sacred, is theirs by God-given right. They complained of attacks and other provocations by Palestinians. On February 25, 1994, a medical doctor raised in the United States, Baruch Goldstein, used an assault rifle to kill more than thirty Muslim worshippers in a mosque in Hebron on the West Bank. He was overwhelmed by the crowd and beaten to death.

In 2003 Juergensmeyer brought out a third, "completely revised" edition that looks at the events of 2001 and shows how they were part of the same general trend.[47] He summarizes the part played by religion as follows:

> Religion is crucial for these acts, since it gives moral justifications for killing and provides images of cosmic war that allow activists to believe they are waging spiritual scenarios. This does not mean that religion causes violence, nor does it mean that religious violence cannot, in some cases, be justified by other means. But it does mean that religion often provides the mores and symbols that make possible bloodshed—even catastrophic acts of terrorism.[48]

NOTES

1. Jonathan Kirsch, *God against the Gods: The History of the War between Monotheism and Polytheism* (New York: Viking Compass, 2004), from book jacket.

2. Ibid., p. 2.

3. Monotheism did not start with the Jews. The first major monotheist was the fourteenth-century BCE Egyptian pharoah Akhenaton. Not until the seventh century BCE under the reign of King Josiah was Israel purged of polytheistic beliefs.

4. In this section I have relied heavily on *The Dark Bible* located at http://www.nobeliefs.com/DarkBible/DarkBibleContents.htm (accessed December 13, 2008), produced by nobeliefs.com, version 2.09, September 20, 2008.

5. For selected excerpts of cruelty and violence in the Qur'an, see *The Skeptic's Annotated Qur'an,* http://skepticsannotatedbible.com/Zquran/cruelty/short.html (accessed December 31, 2008); for a history of the Qur'an, see Ibn Warraq, ed., *The Origins of the Koran: Classic Essays on Islam's Holy Book* (Amherst, NY: Prometheus Books, 1998).

6. Dan Barker, *Godless: How an Evangelical Preacher Became One of America's Leading Atheists* (Berkeley, CA: Ulysses Press, 2008), pp. 178–83.

7. James A. Haught, *Holy Horrors: An Illustrated History of Religious Murder and Madness*, updated ed. (Amherst, NY: Prometheus Books, 2002).

8. James A. Haught, "Holy Horrors," *Penthouse*, August 1990, reprinted in James A. Haught, *Honest Doubt: Essays on Atheism in a Believing Society* (Amherst, NY: Prometheus Books, 2007), pp. 61–72.

9. Vox Day, *The Irrational Atheist: Dissecting the Unholy Trinity of Dawkins, Harris, and Hitchens* (Dallas [Chicago]: BenBella Books distributed by Independent Publishers Group, 2008).

10. Ibid., p. 224.

11. Ibid., p. 213.

12. Sam Harris, *The End of Faith* (New York: Norton, 2004), p. 101; Richard Dawkins, *The God Delusion* (Boston: Houghton Mifflin, 2006), p. 274.

13. Peter De Rosa, *Vicars of Christ: The Dark Side of the Papacy* (New York: Crown, 1988), pp. 6–7. Page references are to the 1991 Gorgi pbk. ed.

14. Dawkins, *The God Delusion*, p. 273.

15. Day, *The Irrational Atheist*, p. 238.

16. Ibid., pp. 238–39.

17. Some scholars believe that figure was greatly inflated by cold war propaganda.

18. De Rosa, *Vicars of Christ.*

19. Ibid., p. 107.

20. Ibid., p. 126.

21. Hector Avalos, *Fighting Words: The Origins of Religious Violence* (Amherst, NY: Prometheus Books, 2005), pp. 326–31.

22. Edvard Radzinsky, *Stalin: The First In-Depth Biography Based on Explosive New Documents from Russia's Secret Archives* (New York: Anchor, 1997).

23. Avalos, *Fighting Words,* p. 330.

24. Jon Krakauer, *Under the Banner of Heaven: A Story of Violent Faith*, 1st Anchor Books ed. (New York: Anchor Books, 2004). Originally published in 2003.

25. Paul Kurtz, *The Transcendental Temptation: A Critique of Religion and the Paranormal* (Amherst, NY: Prometheus Books, 1986) and references therein. See also Wikipedia and other sources on the Internet.

26. Krakauer, *Under the Banner of Heaven*, p. 79.

27. Ibid., p. 188.

28. Ibid., p. 219.

29. Ibid., p. 220.

30. Ibid., p. 227.

31. Susan Greene, "Polygamy Prevails in Remote Arizona Town," *Denver Post,* June 4, 2001, online at http://www.rickross.com/reference/polygamy/polygamy 54.html (accessed December 15, 2008).

32. Krakauer, *Under the Banner of Heaven*, p. 82.

33. Ibid., p. 166.

34. Ibid., p. 188.

35. Ibid., pp. 301–302.

36. Ibid., pp. 306–307. The symptoms are listed here.

37. Ibid., pp. xviii–xix.

38. Taner Edis, *An Illusion of Harmony: Science and Religion in Islam* (Amherst, NY: Prometheus Books, 2007).

39. Taner Edis, "A False Quest for a True Islam," *Free Inquiry* 25, no. 5 (2007): 48–50.

40. Bruce Lincoln, *Holy Terrors: Thinking about Religion after September 11* (Chicago: University of Chicago Press, 2003).

41. Ibid., p. 16.

42. Ibid., pp. 97–104.

43. Ibid., p. 100.

44. Ibid., p. 102.

45. Mark Juergensmeyer, *Terror in the Mind of God: The Global Rise of Religious Violence* (Berkeley: University of California Press, 2000).

46. Ibid., pp. 24–30.

47. Mark Juergensmeyer, *Terror in the Mind of God: The Global Rise of Religious Violence*, 3rd rev. and updated ed. (Berkeley: University of California Press, 2003).

48. Ibid., p. xi.

SUFFERING AND MORALITY

Is God willing to prevent evil, but not able? Then he is impotent.
Is he able but not willing? Then he is malevolent.
Is he both able and willing? Whence, then evil?
—David Hume, attributed to Epicurus[1]

A CRISIS IN FAITH

Bart Ehrman is a reputable biblical scholar who is the author of three best sellers on the New Testament: *Misquoting Jesus: The Story behind Who Changed the Bible and Why* (2005), *God's Problem: How the Bible Fails to Answer Our Most Important Question—Why We Suffer* (2008), and *Jesus, Interrupted: Revealing the Hidden Contradictions in the Bible (and Why We Don't Know about Them)* (2009).[2] I have read these and listened to the audio of his course on *The Historical Jesus*.[3] For most of his life, Ehrman was a committed Christian. In high school he was "born again," but later he wondered what he had converted from since he had always gone to church, prayed to God, and confessed his sins. He trained for the ministry at Moody Bible

Institute, a fundamentalist college in Chicago, finishing his undergraduate work at Wheaton, an evangelical Christian college in Illinois (Billy Graham's alma mater). Ehrman then went on to graduate school at Princeton Theological Seminary, where he specialized in studying the New Testament in its original Greek and received a master of divinity, which qualified him as a minister, and a PhD in New Testament studies.

During these years he was involved in several churches, serving as a youth pastor. After graduating from seminary he served as a pastor at American Baptist Church and Princeton Baptist Church.

But then, he explains, "I started to lose my faith. I now have lost it altogether. I no longer go to church, no longer believe, no longer consider myself a Christian."[4] His book explains why. It's a long story, but here is the short version:

> I realized that I could no longer reconcile the claims of faith with the facts of life. In particular, I could no longer explain how there can be a good and all-powerful God actively involved in this world, given the state of things. For many people who inhabit this planet, life is a cesspool of misery and suffering. I came to a point where I simply could not believe that there is a good and kindly disposed Ruler who is in charge of it.[5]

Ehrman relates how at a Christmas Eve service he listened to a layman read from the Book of Common Prayer. The refrain repeated several times is: "You came into the darkness and made a difference. Come into the darkness again." Ehrman tells how he sat there in tears, listening:

> But these were not tears of joy. They were tears of frustration. If God had come into the darkness with the advent of the Christ child, bringing salvation to the world, why is the world in such a state? Why *doesn't* he enter the darkness again? Where is the presence of God in this world of pain and misery? Why is the darkness so overwhelming?[6]

Ehrman responds to some of the answers theists give to justify God's allowing suffering. For example,

> One answer... is that [God] intervenes in the hearts of the suffering, bringing them solace and hope in the time of their darkest need. It's a nice thought, but I'm afraid that from where I sit, it simply isn't true. The

> vast majority of people dying of starvation, or malaria, or AIDS feel no solace or hope at all—only sheer physical agony, personal abandonment, and mental anguish.[7]

Note that the author is using scientific reasoning. "Look around!" he exclaims. Don't just try to think of some comforting explanation. Look at the data!

God's Problem addresses the various ways the Bible has attempted to address the problem of suffering, which, after all, has always been part of human life. In what follows I will briefly summarize these, but this will be no substitute for reading Ehrman's passionate book.

THE PROPHETS' ANSWER

Much of the early Old Testament is concerned with the prophets, who taught that the sufferings of the Jews—lost wars, pestilence, famine, exile—were God's punishment for their sins.

The book of Amos tells the people of Israel that they will be destroyed for breaking the law of God. He was particularly concerned about social injustice—the wealthy oppressing the poor. Since the Jews were the chosen people, their punishments were particularly severe:

> You only have I known of all the families of the earth; therefore I will punish you for all your iniquities. (Amos 3:2)[8]

And here's how:

> An adversary shall surround the land and strip you of your defense: and your strongholds shall be plundered. (Amos 3:11)

Amos was from the south. Hosea was a contemporary of Amos from the north who was dismayed that the people of Israel had begun worshipping other gods, such as Baal:

> I will destroy you, O Israel;
> who can help you? (Hosea 13:9)

Because the chosen people rebelled against God, Hosea predicted that the capital of the northern kingdom, Samaria, would bear the guilt:

> They shall fall by the sword,
> Their little ones shall be dashed in pieces,
> and their pregnant women ripped open. (Hosea 13:16)

Ehrman comments, "This is not the kind, loving, caring God of nursery rhymes and Sunday school booklets. God is a fierce animal who will rip his people to shreds for failing to worship him."[9]

Shortly after, Assyria marched against Israel, destroyed its capital of Samaria, as Hosea had predicted, and sent many of the people into exile.

Ehrman quotes extensively from these and two other better-known prophets who carried similar messages: Isaiah and Jeremiah. He cautions that they were all speaking to the people of their times and people today should not read anything into it that relates to contemporary life.

The prophets were not the only ones to preach that God makes people suffer. This notion can be found in the historical books, Psalms, and Proverbs. Ehrman asks:

> Why do people suffer? In part, it is because God makes them suffer. It is not that he merely causes a little discomfort now and then to remind people that they need to pay more attention to him. He brings famine, drought, pestilence, war, and destruction. Why do God's people starve? Why do they incur dreadful and fatal diseases? Why are young men maimed and killed in battle? Why are entire cities laid under siege, enslaved, destroyed? Why are pregnant women ripped open and children dashed against rocks? To some extent, at least it is God who does it. He is punishing people when they have gone astray.[10]

There is a threat here, but also a promise. The promise is also clearly stated in the book of Proverbs:

> Whoever is steadfast in
> righteousness will live,
> But whoever pursues evil will
> die. (11:19)

> No harm happens to the
> righteous,
> but the wicked are filled with
> trouble. (12:21)

Now, these are predictions that can be tested scientifically. Are only the wicked punished while the righteous are free from suffering? Hardly. We can quite confidently say that this prediction of the Bible is falsified by the data. The prophets' solution to the problem of suffering is provably wrong.

FREE WILL AND SUFFERING

One of the most common responses to the problem of suffering is the *free will* defense. As Ehrman puts it: "If God had not given us free will, this would be a less-than-perfect world, but God wanted to create a perfect world, and so we have free will—both to obey and disobey him, both to resolve suffering and to cause it. This is why there is suffering in a world ultimately controlled by a God who is both all powerful and loving."[11] I think the answer to this defense is obvious. Not all suffering is caused by other humans. Much of it happens as the result of natural disasters beyond the control of humanity. Ehrman mentions the volcanic eruption that occurred in northern Colombia, South America, on November 13, 1985. Mudslides killed more than thirty thousand people. Then there was the Indian Ocean tsunami of December 26, 2004, where the death toll is uncertain but somewhere around three hundred thousand. That's one hundred times the human-caused deaths in the Twin Towers on 9/11. Of course, we cannot forget the millions who died from human actions in the last century, and centuries before. But they still are not all who have suffered. In fact, the influenza epidemic of 1918 killed more Americans than all of the wars of the twentieth century put together.[12] And that was not the result of a free choice made by some human.

REDEMPTIVE SUFFERING

So far, then, the Bible proposes that suffering is caused by God as a punishment for sin, disobedience, or by humans acting in their free will. The scientific facts are that being virtuous and totally obedient is no guarantee against suffering and that there still would be plenty of suffering in the absence of human free will.

The next type of suffering Ehrman identifies as present in biblical teachings is *redemptive suffering.* As he explains the concept, "Sometimes God brings good out of evil, a good that would not have been possible if the evil had not existed."[13]

As one example Ehrman gives the biblical story of Joseph, who is sold by his brothers into slavery and ends up in Egypt, where he becomes a successful slave but then spends years in prison on a false charge. His gifts of prophecy lead him to be freed and he becomes the pharaoh's right-hand man. With Israel threatened by starvation, Joseph's brothers appear before him to buy food. He eventually reveals himself to them. They humbly beg forgiveness, and Joseph says, "Am I in the place of God? Even though you intended harm to me, God intended it for good, in order to preserve a numerous people, as he is doing today" (Gen. 50:19–20). The book of Genesis thus ends with God saving his people through Joseph's suffering.

More disturbing is Ehrman's take on the New Testament story of Lazarus, whom Jesus raises from the dead in his most important miracle.[14] Reading the whole story (as few Christians do) we find that Jesus learned of Lazarus's illness days earlier from Lazarus's sisters. He was just a few days' journey away but refused their entreaty to come quickly. Why? Because, Jesus says, this illness "is for God's glory, so that the son of God may be glorified through it" (John 11:4). Apparently Jesus deliberately stays away so that Lazarus is good and dead, buried for four days, and putrefying noticeably, then Jesus raises him from the dead in front of a crowd "so that the Son of God might be glorified."

In these and other passages of Scripture, suffering is experienced so God may be glorified or can otherwise bring good out of it. And, of course, the link between suffering and salvation is manifested in Jesus' crucifixion on the cross. The myth of Jesus as the messiah was expressed in 1 Peter 2:24: "He bore our sins in his body on the cross, so that we might be freed from sins and live for righteousness, for by his wounds you have been healed." These words refer to the famous passage in Isaiah that are sung in the heavenly tones of Handel's great oratorio, *Messiah*:

> Surely he hath borne our griefs, and carried our sorrows: yet we did esteem him stricken, smitten of God, and afflicted. But he was wounded for our transgressions, he was bruised for our iniquities: the chastisement of our peace was upon him; and with his stripes we are healed. (Isa. 53:4–5)

Dashing unholy water on the usual interpretation of these lines as predicting the messiah, Ehrman argues that the prophet Isaiah himself made no mention of the messiah and identified his "suffering servant as the nation of Israel."[15] Jesus was the opposite of what most Jews regarded as the messiah of the Old Testament.

But the basic idea is that without punishment there can be no atonement for sin. Suffering can be redemptive. It can bring about salvation.

Perhaps it can. Let us again adopt the scientific perspective and look at the data. What was the redemptive value of the Crusades or the Black Plague or the Holocaust? What is the redemptive value of one child dying of leukemia or millions of children starving to death? The redemptive value would have to be enormous to justify the huge amount of suffering involved in those events. And, if the purpose of God coming to Earth in the person of Jesus was to suffer himself and thereby redeem humanity, then why wasn't that enough? What suffering could be greater than the suffering of God himself? Why, then, does humanity suffer as much now as it did before Christ?

THE BOOK OF JOB

The book of Job tells the story of a man who has everything a man could want and at the same time is the most pious of men, obeying every commandment to the letter and performing every required ritual and sacrifice. Satan tells God that Job is only doing this to earn God's favor, so God takes everything away, including all his children, and makes Job the most miserable man on Earth. Still Job does not curse God but continues his obeisance. However, egged on by friends, he asks God to allow him to plead his case before him. God rebukes Job for thinking that he, God, has anything to explain. Job repents and God restores him to his previous condition, with a new batch of children, and Job lives to a ripe old age.

I suppose the moral is supposed to be that one should remain righteous in the face of all adversity and all will be well in the end. But that's not much comfort for the original batch of Job's children who lost everything permanently.

The book of Job is one of the most powerful arguments against God in the atheist arsenal. It proves once and for all that YHWH is not a good god. Even if such a god exists, we surely cannot rely on him to define our morals, to tell us what is good or bad. If holiness is what gods do, then holiness is a terrible thing.

JESUS, THE APOCALYPTIC TEACHER

In *God's Problem*, the excellent lecture course *The Historical Jesus*, and other writings, Bart Ehrman makes the case that Jesus was primarily an apocalyptic teacher. Apocalyptic thinking can be found throughout the Old Testament, notably the book of Daniel, and was widespread and influential at Jesus' time. We saw in chapter 2 that it remains a major part of much of Christianity outside the Catholic Church and mainline Protestantism, as well as other faiths.

There are four tenets of apocalypticism:

1. *Dualism.* Two fundamental components of reality exist, the forces of good controlled by God and the forces of evil controlled by Satan, the devil.
2. *Pessimism.* The forces of evil are in control of the world and there is nothing any of us can do about it until God decides to intervene.
3. *Vindication.* God will intervene, sending a savior from heaven called "the Son of Man," who will bring judgment to Earth.
4. *Imminence.* This judgment will happen very soon.

Jesus is not viewed by many today as an apocalyptic teacher, even among those who believe Judgment Day is coming. This picture of Jesus is generally not taught in Sunday school. However, Ehrman claims that the majority of critical scholars in the English- and German-speaking worlds have understood Jesus as a Jewish apocalypticist for more than a century, since the publication of Albert Schweitzer's *The Quest for the Historical Jesus.*[16]

Whether or not Jesus was himself an apocalypticist (or, indeed, whether or not he even existed), the New Testament is filled with apocalyptic teaching. Indeed, it does not originate with Jesus but with John the Baptist.

However, the Gospels have Jesus carry forth the message of a cosmic figure called the Son of Man[17] coming down from the clouds in judgment and establishing God's kingdom on Earth:

> In those days after the affliction, the sun will grow dark and the moon will not give its light, and the starzs will be falling from heaven, and the powers of the sky will be shaken; and then they will see the Son of Man coming on the clouds with great power and glory. Then he will send forth his angels and he will gather the elect from the four winds, from the end of earth to the end of heaven. (Mark 13:24–27)

And this is the Bible's ultimate answer to suffering. Those who suffer in this world, the hungry and the poor, can look forward to an end of that suffering, indeed an end to death. Those who cause pain and suffering will be punished. Despite the claims of many of today's feel-good preachers, who tell their congregations that God wants them (and, the preacher himself) to be prosperous, the rich will not do so well as Jesus tells it:

> Woe to you who are rich,
> for you have received your
> consolation.
> Woe to you who are full now,
> for you will be hungry. (Luke 6:24–25)

A SCIENTIFIC TAKE

What is a scientist to make of all of this? We have already seen how the promise of the prophets and other biblical writers—that you can avoid suffering by righteous behavior—is falsified by the evidence. The good suffer just as much as the bad. When we look at the "theory" of the apocalypse we see that one of the basic assumptions is *imminence.* That is, the apocalypse is right around the corner. However, as we saw in chapter 2, in four different places in the Gospels Jesus promised his disciples that they would live to see the Second Coming. Certainly that prediction has long since been falsified, as have the many other apocalyptic predictions throughout the centuries. I think we can safely say it will never happen.

ALL IS VANITY

Ehrman's favorite book of the Bible is also mine—Ecclesiastes. The author, or "preacher," claims to be none other than Solomon, son of David and king of Israel and Judea. Scholars think this is unlikely. Nevertheless, the author's advice is, as we will see in future chapters, the wisdom of the ages:

> Live joyfully with the wife whom thou lovest all the days of the life of thy vanity, which he hath given thee under the sun, all the days of thy vanity; for that is thy portion in this life, and in thy labor which thou taketh under the sun. Whatsoever thy hand findeth to do, do it with thy might; for there is no work, nor device, nor knowledge, nor wisdom, in the grave wither thou goest. (Eccles. 9:9–10, KJV)

SUFFERING IN ISLAM

Muslim theologians do not appear to agonize over the problem of suffering the way Christian theologians do. Everything that happens in the world happens by Allah's will. The Qur'an says that Allah is infinitely wise, right, good, fair, and just. Of course that's logically inconsistent, but again that doesn't bother Muslims. They simply assert that finite beings cannot fully grasp his will and must surrender to it without question.

The best the Muslims can conclude based on the Qur'an is that suffering occurs to teach them that they must adhere to Allah's natural and moral laws. Sometimes suffering is a punishment or a test.[18]

According to the Qur'an our present life, with all its joys and suffering, is merely transitory and illusionary. Life after life is permanent and perpetual. Many who have lived lives full of suffering may enjoy the most blissful everlasting life in the next world. Many who have enjoyed the sinful pleasures of the material world may go through grievous torment in life yet to come.

Natural calamities, illnesses, and other sufferings occur in consequence of natural laws that are necessary in the vast universe in which human life is a small portion. God the Creator put these laws in place to actually support and evolve life on Earth, and they do not reflect the Creator's cruelty toward humans.[19]

Man-made afflictions are caused by the abuse of free will, without which humans would lose the very essence of their existence. God commands humans not to inflict harm on others, or they will face the consequences.

So it does not matter how much a person suffers in this world, as long as he or she is engaged in doing good and repelling evil.

> The joys and comforts of the life yet to come are far greater, unparalleled, and everlasting as compared to the human sufferings of this life!
>
> These sufferings are a trial, a test from God. If we persevere God will grant us boundless joy, happiness, and pleasure.[20]

Clearly the great appeal of Islam and Christianity is their promise of eternal life. No matter how much you suffer in this world, if you do good and fight evil, or God simply grants it "by grace," the eternal joys and pleasures of the next world will far outweigh the transient worldly pain you had to endure.

But it all depends on the existence of God, an eternal human soul, and an afterlife. As we will see, these can all now be ruled out beyond a reasonable doubt—or at least sufficient doubt to make one think twice before completely submitting his life in this world to the will of God or Allah. You could say, like Blaise Pascal (d. 1662), that you have everything to gain and nothing to lose (*Pascal's wager*); but you do have a lot to lose—your freedom to live life as you choose. You have a lot to gain if Islam is true, but you also have a lot to gain by trying out for National Football League quarterback.

SUFFERING IN HINDUISM

Hinduism's answer to suffering also has to do with the notion of everlasting life. Only here, instead of a heaven of bliss or a hell of agony following life on Earth, Hindus believe that the soul undergoes a continuous cycle of earthly lives (*samsara*), eventually terminating in a state of ultimate peace called *moksha*. This belief has a rational basis, following from the hypothesis of a just and moral universe. Simple (scientific) observation of the world shows no evidence for justice in a single life. The virtuous may suffer, and the vicious may prosper. These attributes of individuals cannot be assigned to the will of God since that would be inconsistent with a loving God. So these differences must be the result of the individual's own actions in a previous life.

Hindus find the notions of heaven and hell as the places where justice is meted out to be unsatisfactory since they are different domains than Earth. The conditions are different. Actions performed on Earth should be rewarded or punished on Earth. This is the *law of karma*.[21]

In writing this I noticed something that I have never seen pointed out in my admittedly limited study of Hinduism. Of course, there is much more to the Hindu religion than what is mentioned above, in particular, the worship of many deities who apparently can act in response to prayers. However, the answer to the problem of suffering I have briefly outlined above could still apply in a purely natural world with no deities or supernatural forces of any kind. Unlike other religions, Hindu gods do not act to reward or punish individuals. The law of karma, which can be assumed to be a natural law along with the hypothesis that the universe is just, governs a person's life based on his past performance. I'm not saying I can write down some equations that model the law of karma naturally, but the point is that no gods need participate.

So the Hindu explanation is a rational one, but it is based on the hypotheses of a just universe and the law of karma. If there were real evidence of past lives, that would verify the Hindu claim. However, not being able to link up a present life with a past one, we cannot verify a connection between behavior in the past life and rewards or punishments in the current one. Some parapsychologists have claimed *past-life regression* can occur under hypnosis, but these reports have all proven to be phony scams.[22]

As with many religious claims, we again find one that is eminently testable by science. All that has to happen is that someone remembers something from her previous life that she could not possible have known and that is verified as correct by other, objective means. That would prove reincarnation. If a large sample could then be collected and a correlation found between suffering or reward and past behavior, then the law of karma would be proved. Note that the rest of Hinduism, especially the existence of gods, would not necessarily follow.

However, this has not happened. With the billions of humans who have died since humans became humans, you would think that some sign of rebirth would have occurred by now. The fact that it has not allows us to conclude that reincarnation is falsified beyond a reasonable doubt.

As it is, we have no reason to believe the Hindu tradition. Like all the other religions of the world, Hindu beliefs are based on faith in the total absence of evidence, evidence that should be there if the religion is true.

SUFFERING IN BUDDHISM

Buddhism is pretty much all about suffering.[23]

> Suffering is a big word in Buddhist thought. It is a key term and it should be thoroughly understood. The Pali word is dukkha, and it does not just mean the agony of the body. It means that deep subtle sense of unsatisfactoriness which is a part of every mind moment and which results directly from the mental treadmill. The essence of life is suffering, said the Buddha. At first glance this seems exceedingly morbid and pessimistic. It even seems untrue. After all, there are plenty of times when we are happy, aren't there? No, there are not. It just seems that way. Take any moment when you feel really fulfilled and examine it closely. Down under the joy, you will find that subtle, all-pervasive undercurrent of tension, that no matter how great this moment is, it is going to end. No matter how much you just gained, you are either going to lose some of it or spend the rest of your days guarding what you have got and scheming how to get more. And in the end, you are going to die. In the end, you lose everything. It is all transitory.[24]

According to Buddhism, suffering is all in the mind, caused by "root delusions" such as attachment, anger, and ignorance. These delusions cause us to take negative actions and create negative experiences.

Since Buddhists believe in reincarnation, suffering does not end with death but continues. To end the cycle of pain one must enter the state of *nirvana*, that is, beyond suffering. We can reach nirvana by ridding ourselves of our delusions.

The way to do this is called the *eightfold path to enlightenment* by which we control our body and mind, helping others instead of doing harm:

The Eightfold Path to Enlightenment

1. **Correct thought:** avoiding covetousness, the wish to harm others, and wrong views (like thinking actions have no consequences, I never have any problems, there are no ways to end suffering, etc.)
2. **Correct speech:** avoid lying, divisive and harsh speech, and idle gossip
3. **Correct actions:** avoid killing, stealing, and sexual misconduct
4. **Correct livelihood:** try to make a living with the above attitude of thought, speech, and actions
5. **Correct understanding:** developing genuine wisdom
6. **Correct effort:** after the first real step we need joyful perseverance to continue
7. **Correct mindfulness:** try to be aware of the "here and now," instead of dreaming in the "there and then"
8. **Correct concentration:** to keep a steady, calm, and attentive state of mind

(The last three aspects refer mainly to the practice of meditation.)[25]

Again it is important to note that, although many Buddhists throughout history have believed in gods and the supernatural, these particular teachings of Buddha do not depend on the existence of anything beyond the material world.

Reaching nirvana ends the cycle of rebirth, and if there is no heaven or hell, death ends in nothingness. As I mentioned in the discussion of Hinduism above, reincarnation is empirically testable and has failed these tests for a sufficiently long time to conclude it is falsified beyond a reasonable doubt. It follows that we have one life that ends in nothingness.

It remains to be seen whether nirvana can be achieved before death. It seems worth a try. In a future chapter we will examine the possibility of a "Way of Nature" by which an atheist can live a happy, fulfilled life and approach death with peaceful acceptance.

SUFFERING IN TAOISM

While Taoism does not put the same emphasis on suffering as Buddhism, its solution is similar to what we have been hearing. Here is a verse from the Tao Te Ching:

Favor and disgrace make one fearful
The greatest misfortune is the self
What does "favor and disgrace make one fearful" mean?
Favor is high; disgrace is low
Having it makes one fearful
Losing it makes one fearful
This is "favor and disgrace make one fearful"

What does "the greatest misfortune is the self" mean?
The reason I have great misfortune
Is that I have the self
If I have no self
What misfortune do I have?

So one who values the self as the world
Can be given the world
One who loves the self as the world
Can be entrusted with the world[26]

So while the source of misfortune is the self, the Tao does not require us to completely reject ego. However, we should not focus on it too much and neglect the rest of the world.

RELIGION AND MORALITY

Every religion has as a central practice, a set of moral codes to which its members are expected to adhere. So, there can be no doubt that morality and religion are closely related. Indeed, most people have been brain-washed by priests and preachers to think that religion is necessary for morality—that if it weren't for religion everybody would do whatever he or she wanted and commit all kinds of heinous acts. In every religion these rules of behavior are seen as arising from divine rather than human origin. This gives these principles special meaning. They are not to be questioned. They are worth dying for. They are worth killing for.

But the scientific facts—objective observations of human behavior—as well as logical analysis tell a different story. In an essay titled "Morality without Religion," philosophers Marc Hauser and Peter Singer elucidate the problems associated with the view that morality comes from God.[27] First, it is a tautology to say that God is good and defines for us what is good and bad. All that says is that God is good according to his own standards. As Plato, using the "voice" of Socrates pointed out, either God defines what is good, in which case it is arbitrary, or God is inherently good, in which case goodness is defined independent of God.[28]

Second, there are no moral principles that are shared by all religious people, independent of their affiliation, that are not also shared by atheists and agnostics. Hauser and Singer describe studies in which subjects are asked to resolve certain moral dilemmas. The results show no statistically significant differences between religious people and atheists and agnostics. Regardless of whether they are right or wrong according to some standard, they think the same way. If, as preachers say, we need religion to keep us from doing heinous acts, why don't all atheists and agnostics—not just Stalin and Pol Pot—perform heinous acts? And why do so many believers perform so many heinous acts?

As we saw in chapter 5, believers try to argue that atheism is more evil than theism because twentieth-century atheists such as Hitler, Stalin, and Pol Pot killed more people than the Crusaders, the Inquisitors, and the kings of Christendom put together. I showed that this statement is disputable since Hitler was not an atheist and the numbers themselves are arguable.

But let's look at it in a different way. Stalin and his comrades had more power and more population to work with. How do you measure evil, anyway? Is the evil of one individual torturing one child to death somehow less than killing a million people? Child and spouse abuse is highest among the most fundamentalist Christians and Muslims.[29] If you went by percentages alone, today and throughout history, you would conclude that evil is committed by a far higher percentage of believers than nonbelievers. And even among the religious, you will find evil committed by a far higher percentage of god believers than among Taoists, Buddhists, Jains, and other religions that do not view morality as god-given.

A third difficulty for the view that morality has its origin in religion is that despite the huge doctrinal difference between religions and wide differences between cultures, all religions along with atheists and agnostics agree on some elements of morality that seem to be universal. In *The Evolution of Morality and Religion*, biologist Donald L. Broom shows that the great diversity of religions have common moral codes. They differ only in "more peripheral" aspects that may be conspicuous and regarded by believers as of central importance but are really secondary.[30]

UNIVERSAL MORALITY

The most important of these common moral codes is usually called the *Golden Rule*, but it really includes more than the familiar homily "Do unto others as you would have them do unto you." The great religions have expressed this notion in a variety of ways that are generally more demanding.[31] A better representation of a universal moral code is *altruism*, where you are not only to treat others as you would want to be treated yourself, but to be willing to treat them better—to sacrifice your personal welfare and even your life for that of others.

Let me review statements of the universal moral code from various religions, more or less chronologically. In the Hindu Mahabharata (c. 2000 BCE), Brihaspati says:

> One should never do that to another which one regards as injurious to one's own self. This, in brief, is the rule of dharma. Other behavior is due to selfish desires.[32]

The entire philosophy of Jainism (date of origin unknown) is based on avoiding the suffering of any living thing, a true altruism much more demanding than the Golden Rule:

> Just as sorrow or pain is not desirable to you, so it is to all which breathe, exist, live or have any essence of life. To you and all, it is undesirable, and painful, and repugnant.[33]

In the Old Testament, Leviticus (written c. 1400 BCE) tells you to love your neighbor as yourself:

> Thou shalt not avenge, nor bear any grudge against the children of thy people, but thou shalt love thy neighbor as thyself. (Lev. 19:18, KJV)

Usually you hear just the last phase. However, the full verse makes it clear Leviticus is talking about other members of your tribe, not all people. The same is true for the commandment "Thou shall not kill."

The Buddha (c. 500 BCE) makes it clear that our own happiness is at stake in being altruistic:

> One who, while himself seeking happiness, oppresses with violence other beings who desire happiness, will not attain happiness hereafter. (Dhammapada 10)

The Tao Te Ching (c. 500 BCE) clearly teaches we should put others ahead of ourselves:

> The sage has no interest of his own, but takes the interests of the people as his own. He is kind to the kind; he is also kind to the unkind: for Virtue is kind. He is faithful to the faithful; he is also faithful to the unfaithful: for Virtue is faithful. (Tao Te Ching, v. 49)

Confucius (c. 500 BCE) gave the common version of the Golden Rule: "Do not impose on others what you do not wish for yourself."[34] Recall it said in Leviticus to "Love thy neighbor as thyself," but the verse is clear in that it is talking about others in your tribe. In the Sermon on the Mount, Jesus (c. 30 CE) extends this to your enemies: "But I say unto you which hear, Love your enemies, do good to them which hate you, bless them that curse you, and pray for them which despitefully use you" (Luke 6:27–28, KJV). This may be perhaps the only original moral teaching in the Gospels, but it is similar in concept to the Tao Te Ching verse given above. Jesus also gives his version of the Golden Rule: "And as ye would that men should do to you, do yet also to them likewise" (Luke 19:31, KJV).

Most scholars insist there is no Golden Rule in Islam, where the faithful are commanded to treat all Muslims as brothers but generally regard killing infidels as a minor offense. However, Muhammad (600 CE) supposedly did say, "That which you want for yourself, seek for mankind."[35] Perhaps he was thinking of the day when all mankind would be Muslim.

I think the fact that the moral principle of working to minimize the suffering of others arises in all religions, including those without gods, attests to its human rather than divine origin.

THE NATURAL ORIGIN OF MORALITY

Hauser and Singer summarize the conclusion of the laboratory studies of morality as follows:

> These studies begin to provide empirical support for the idea that like other psychological faculties of the mind, including language and mathematics, we are endowed with a moral faculty that guides our intuitive judgments of right and wrong, interacting in interesting ways with the local culture. These intuitions reflect the outcome of millions of years in which our ancestors have lived as social mammals, and are part of our common inheritance, as much as our opposable thumbs are. These facts are incompatible with the story of divine creation.[36]

They continue:

> Our evolved intuitions do not necessarily give us the right or consistent answers to moral dilemmas. What was good for our ancestors may not be good for human beings as a whole today, let alone for our planet and all the other beings living on it. But insights into the changing moral landscape (e.g., animal rights, abortion, euthanasia, international aid) have not come from religion, but from careful reflection on humanity and what we consider a life well lived. In this respect, it is important for us to be aware of the universal set of moral intuitions so that we can reflect on them and, if we choose, act contrary to them. We can do this without blasphemy, because it is our own nature, not God, that is the source of our species' morality. Hopefully, governments that equate morality with religion are listening.[37]

Broom argues that human codes of conduct evolved as a consequence of natural selection. He shows that social behavior even among animals is *not* a "largely uninterrupted competition in which those who seek their own immediate advantage are the most successful."[38] He gives the examples of birds not fighting over territory with neighboring rivals and chimpanzees not forcibly taking food from weaker members of their clans.

Scientific American columnist and *Skeptic* magazine publisher Michael Shermer, in his important work *The Science of Good and Evil*, lists other examples of animal altruism.[39] He also gives some additional examples of the Golden Rule.[40]

Shermer gives a plausible but largely speculative account of the natural evolution of morality, which is currently a subject of continuing research and a rapidly growing body of literature. In *Breaking the Spell*, Daniel C. Dennett reviews the conclusions of his own thinking and that of several other authors who have begun to make substantial progress in understanding how religion and morality evolved.[41]

In *Darwin's Cathedral: Evolution, Religion, and the Nature of Society*, evolutionary biologist David Sloan Wilson hypothesizes that religion is a social phenomenon that evolved by a process of *group selection* to improve cooperation among human groups.[42] As Dennett emphasizes, Wilson's idea of group selection has been met with considerable skepticism within the biological and the anthropological communities. I see two problems: (1) no mechanism is provided, such as we have in genetic selection in biology; (2) religions have been as much a source of group warfare as group cooperation.

More agreeable to most observers are the ideas of anthropologists Pascal Boyer and Scott Atran, who are also trained in evolutionary theory and cognitive psychology.[43] They argue that to understand the hold of religious ideas we need to understand the evolution of the human mind. As with the rest of the body, the brain evolved according to no preconceived plan but developed a patchwork of parts that worked together to improve the reproductive success of our ancestors.

Atran does not see religion as an evolutionary adaptation. Rather, it appears along certain paths in the evolutionary landscape that contain the sorts of social relations and individual emotions that favor religion, but it did not evolve specifically for that purpose.

It seems rather obvious to me that as humans began to live closer to one another they were forced to develop codes of behavior beyond those necessary in a family or tribal setting. Ancient life was violent enough, and if everybody lied, robbed, and killed with abandon, there would be few people left. Those who were left would have a highly dysfunctional society.

I mentioned in chapter 1 how the selfish gene idea promoted by

Richard Dawkins offered an evolutionary explanation for altruism. What is important in evolution is gene survival rather than individual survival. Most parents would trade their lives for that of their child's. Of course they do not intellectualize it this way, but sacrificing themselves offers the better chance that their genes will survive.

NOTES

1. David Hume, *Dialogues concerning Natural Religion*, ed. J. M. Bell (London: Penguin, 1990).

2. Bart D. Ehrman, *Misquoting Jesus: The Story behind Who Changed the Bible and Why* (New York: HarperSanFrancisco, 2005); Bart D. Ehrman, *God's Problem: How the Bible Fails to Answer Our Most Important Question—Why We Suffer* (New York: HarperOne, 2008); Bart D. Ehrman, *Jesus, Interrupted: Revealing the Hidden Contradictions in the Bible (and Why We Don't Know about Them)* (New York: HarperOne, 2009).

3. Bart D. Ehrman, *The Historical Jesus*, audio book (Chantilly, VA: Teaching Company, 2000).

4. Ibid., pp. 2–3.

5. Ibid., p. 3.

6. Ibid., p. 5.

7. Ibid., p. 16.

8. The biblical quotations in this section are taken from Ehrman. Unless otherwise specified, the translations from the Hebrew Bible are from the New Revised Standard version; translations from the New Testament are his own.

9. Ibid., p. 46.

10. Ibid., p. 53.

11. Ibid., pp. 120–21.

12. See http://www.infoplease.com/ipa/A0004615.html and http://www.pandemicflu.gov/general/historicaloverview.html.

13. Ibid., p. 131.

14. Ibid., pp. 136–37.

15. Ibid., p. 139.

16. Ehrman, *God's Problem*, p. 219.

17. In the Gospel narratives, the authors imply that Jesus himself is the "Son of Man." However, Ehrman and other scholars have argued that the historical Jesus anticipated someone else from heaven coming in judgment. See Ehrman, *God's Problem*, p. 283, note 6.

18. Muzammil Siddiqi, "Why Does Allah Allow Suffering and Evil in the

World?" Fatwa Muslim Belief, http://en.allexperts.com/q/Islam-947/Islam-Explain-Suffering.htm (accessed January 15, 2009).

19. This is not to imply that Islam supports Darwinian evolution. God is always fully in control.

20. Mubasher Ahmad, "God and Human Suffering," Islamic Research International, paper presented at an interfaith symposium in Zion City, Illinois, December 8, 2002, http://www.irfi.org/articles/articles_101_150/god_human_suffering.htm (accessed January 15, 2009).

21. Swami Adiswarananda, "Hinduism: The Problem of Suffering—Rebirth and the Law of Karma," Ramakrishna-Vivekananda Center, New York, http://www.ramakrishna.org/activities/message/weekly_message41.htm (accessed January 16, 2009).

22. Victor J. Stenger, *Physics and Psychics: The Search for a World beyond the Senses* (Amherst, NY: Prometheus Books, 1990), pp. 112–13 and references therein.

23. "The Four Noble Truths," *A View on Buddhism*, http://buddhism.kalachakranet.org/4_noble_truths.html (accessed January 17, 2009).

24. Henepola Gunaratana, *Mindfulness in Plain English* (Boston: Wisdom Publications, 2002).

25. "The Four Nobel Truths."

26. Derek Lin, trans. and ed., *Tao Te Ching: Annotated & Explained* (Woodstock, VT: SkyLight Paths, 2006), v. 13.

27. Marc Hauser and Peter Singer, "Morality without Religion," *Project Syndicate*, May 17, 2006, http://www.wjh.harvard.edu/~mnkylab/publications/recent/HauserSingerMoralRelig05.pdf (accessed January 24, 2009).

28. See my discussion of the Euthyphro dilemma and other arguments concerning evil and morality, which I have not repeated in this book, in chapters 7 and 8 of *God: The Failed Hypothesis* (Amherst, NY: Prometheus Books, 2007).

29. Kimberley Blaker, *The Fundamentals of Extremism: The Christian Right in America* (New Boston, MI: New Boston Books, 2003).

30. Donald M. Broom, *The Evolution of Morality and Religion* (Cambridge: Cambridge University Press, 2003), p. 164.

31. See "Ethic of reciprocity," Wikipedia, http://en.wikipedia.org/wiki/Ethic_of_reciprocity#cite_note-27 (accessed January 20, 2009).

32. Mahabharata book 13, Anusasana Parva, section 113, v. 8, http://www.mahabharataonline.com/translation/mahabharata_13b078.php (accessed January 20, 2009).

33. Hermann Jacobi, trans., "Jaina Sutras, Part I," *Sacred Books of the East, Vol. 22* (1884), sutra 155–56.

34. Confucius, Analects XV.24, trans. David Hinton.

35. Sukhanan-i-Muhammad, *The Conversations of Muhammad* (Tehran: Wattles, 1928), p. 192.

36. Hauser and Singer, "Morality without Religion," pp. 3–4.

37. Ibid., p. 4.

38. Broom, *The Evolution of Morality and Religion*, p. 74.

39. Michael Shermer, *The Science of Good and Evil: Why People Cheat, Gossip, Care, Share, and Follow the Golden Rule* (New York: Times Books, 2004), pp. 27–29.

40. Ibid., p. 25.

41. Daniel C. Dennett, *Breaking the Spell: Religion as a Natural Phenomenon* (New York: Viking, 2006), p. 104ff.

42. David Sloan Wilson, *Darwin's Cathedral: Evolution, Religion, and the Nature of Society* (Chicago: University of Chicago Press, 2002).

43. Pascal Boyer, *Religion Explained: The Evolutionary Origins of Religious Thought* (New York: Basic Books, 2001); Scott Atran, *In Gods We Trust: The Evolutionary Landscape of Religion* (Oxford: Oxford University Press, 2002).

7. THE NATURE OF NATURE

Nature made us—nature did it all—not the gods of the religions. Religion is all bunk and all Bibles are man-made.

—Thomas Alva Edison

SCIENTIFIC NATURALISM

We have seen that science is fully capable of detecting the presence of a benevolent god who plays an active role in the universe. So far, it has not done so. Furthermore, a strong case can be made that such a god should have been detected by now, so that absence of evidence can be taken as evidence of absence.

This still leaves open the possibility of a deist god who just set everything in motion but plays no further role, or an evil god who deliberately hides from us and is morally responsible for all the suffering in the world. We have also not ruled out the possibility, widely believed by many, that there still exists some nonpersonal, non-godlike entity—an "ultimate universal" spirit that exists beyond nature and is responsible for bringing the universe and its laws into being.[1] All of these possibilities can be collected under the term *supernaturalism.*

Theologian John Haught gives a reasonable definition of the alternative, which he calls *scientific naturalism*:

1. Apart from nature, which includes human beings and our cultural creations, there is nothing. There is no God, no soul, and no life beyond death.
2. Nature is self-originating, not the creation of God.
3. The universe has no overall point or purpose, although individual human lives can be lived purposely.
4. Since God does not exist, all explanations, all causes are purely neutral and can be understood only by science.
5. All the various features of living beings, including human intelligence and behavior, can be explained ultimately in purely natural terms, and today this usually means in evolutionary, specifically Darwinian, terms.

To these tenets of scientific naturalism, the new atheists would add the following:

6. Faith in God is the cause of innumerable evils and should be rejected on moral grounds.
7. Morality does not require belief in God, and people behave better without faith than with it.[2]

I think this is a good summary of the new atheist position on the nature of the world, with a few minor exceptions. I see no reason to insist that *only* science can understand causes and explanations. And, while philosophers identify different forms of naturalism, most naturalists including the new atheists subscribe to scientific naturalism. Also, nature is not self-originating if it always existed. When I speak of naturalism without any modifier, the reader can safely assume I am referring to scientific naturalism.

I would also add that, according to our best current knowledge, the substance of the universe is matter and nothing else. So, until something other than matter is discovered, always at least a remote possibility, we can equate naturalism with *materialism.*

Furthermore, it is important to note that the statements above defining naturalism are not dogma because science and atheism must never be dogmatic. That is not to say that you won't find individual scientists and atheists who are dogmatic, but the very essence of the scientific method eschews dogmatism. And atheism strives to rely primarily on scientific method, where it can be applied.

Many of the opponents of naturalism are not willing to give naturalists this courtesy. For example, paranormal researcher Charles Tart falsely charges: "They [materialists] don't recognize that their *belief* that everything can be explained in purely material terms should be treated like any scientific theory, i.e., it should be subject to continual test and modified or rejected when found wanting."[3] I cannot think of a single materialist I know who doesn't think materialism is eminently testable and falsifiable, as is any proper scientific theory. In fact, Tart himself would have succeeded in falsifying materialism if the published claim he made about an experiment he performed in 1968 had been independently verified. I told the story in my 1995 book, *Physics and Psychics*,[4] but it is such a beautiful example of how easy it is to be fooled, especially when you want to be fooled, that it bears repeating.

Tart's experiment involved a young woman lying flat on her back on a table having an *out-of-body experience* (OBE). Tart reported that she claimed to float above her body and was able to successfully read a five-digit number on a shelf above her head.[5]

However, two skeptics pointed out that the numbers were visible from the table by the reflection from a wall clock, which Tart did not have the good sense to cover up.[6]

In *Physics and Psychics* I discuss another of Tart's flawed experiments where he claimed to prove psychic effects.[7] I will have more to say about so-called psychic phenomena in chapter 8.

In any case, the question of whether naturalism, as defined above, corresponds to reality must be recognized as provisional. It is based on our best current knowledge. As we have seen, science does not have its mind closed to the supernatural, as supernaturalists and some misinformed scientists continue to erroneously charge. Science has looked and simply sees no evidence for anything beyond nature. Scientists find no empirical or theoretical reason to introduce other-than-natural causes into their models. This includes many areas where supernaturalists insist on supernatural causes, such as the origin of the universe and its laws, the apparent fine-tuning of the constants of physics, the complexity of nature—especially life—and mental phenomena such as self-awareness and cognition. While science does not have a "proven" natural explanation for every one of these situations, we have plausible provisional scenarios that do not require any supernatural component.

Philosopher Michael Rea has argued that naturalism is only a "research program" and cannot be called a "worldview."[8] I don't like to waste my time arguing over the meaning of words. Fortunately, I don't have to here since Richard Carrier has provided a complete rebuttal to Rea's argument.[9] But even if Rea were right, how would that change anything? What's wrong with a research program anyway? It admits the possibility that it may be wrong, which scientists and atheists are perfectly willing to accept, as I hope I have already made very clear.

MATTER

I have a simple definition of *matter*. It's stuff that kicks back when you kick it. The measurable quantity of matter of a body is called *mass*. The more massive a body, the bigger kick you get back when you kick it. When a body is moving, that motion can be described by a quantity of motion called *momentum*, which is for most purposes the product of the mass and the velocity of the body. It is a vector whose direction equals that of the *velocity vector*. The magnitude (length) of the velocity vector is called *speed*. The magnitude of the momentum vector is, for most purposes, the product of mass and speed. At speeds near the speed of light a more complicated formula for momentum, discovered by Einstein and not needed here, must be used.

Both mass and momentum serve to quantify the observed property of matter we call *inertia*. The more massive a body, the harder it is to get moving or stop if it is moving. The higher momentum a body has, the harder it is to change its motion. That is, a greater *force* is needed in these cases.

Another familiar measure of motion is *energy*. It is not independent of mass and momentum. The three are simply related, so by knowing these two you can calculate the third. The three quantities can be thought of as forming a right triangle in which the length of one side is mass, another side is the magnitude of momentum, and the hypotenuse is the energy. Then the Pythagorean theorem can be used to relate the three, as illustrated in figure 7.1. In units where the speed of light $c = 1$, all three quantities have the same units.[10]

Fig. 7.1. The relationship between the energy *E*, momentum *p*, and mass *m* of a body is given by the Pythagorean theorem.

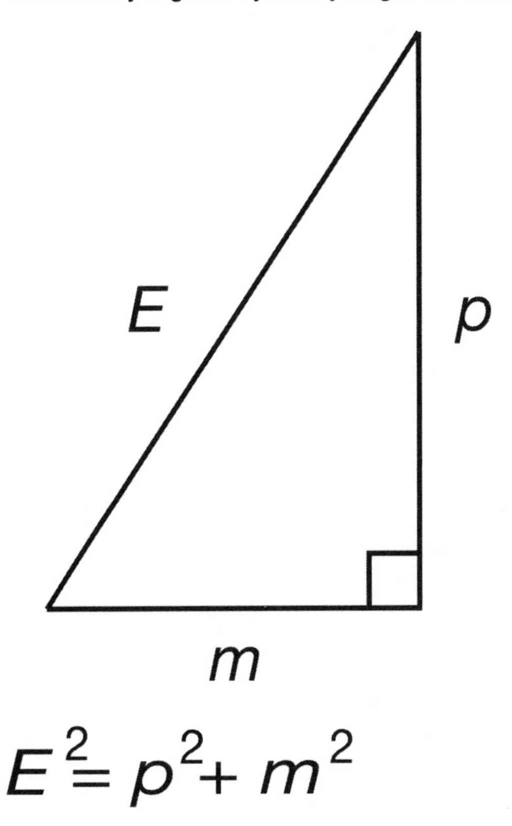

Note that when a body is at rest, its momentum is zero and its energy equals its mass. That is called the *rest energy*. This is the famous $E = mc^2$ relation of Einstein, where by setting $c = 1$ we see more clearly the equivalence of rest energy and mass. When a body is moving, its energy is greater than its rest energy and that added difference is called *kinetic energy*, or energy of motion. The *potential energy* of a body is stored energy that can be positive or negative. For example, a rock held in your hand above the ground has potential energy that is converted into kinetic energy as the rock drops to the ground.

There are actually two kinds of momentum: *linear* momentum, which was described above and applies to rectilinear motion, and *angular* momentum, which applies to rotational motion. When I use "momentum" without a modifier, I will be referring to linear momentum.

My reason for introducing this much detail is the importance of understanding mass, momentum, and energy if you are going to grasp the interplay between physics and theology. The three most important laws of physics are *conservation of linear momentum, conservation of angular momentum,* and *conservation of energy* in which the total linear momentum, total angular momentum, and total energy of an isolated system, respectively, are *conserved*, that is, do not change with time.

A familiar example of linear momentum conservation occurs when firing a gun. The momentum of the bullet is balanced by an equal momentum in the opposite direction that is experienced as a recoil.

Angular momentum conservation is what keeps a spinning top from falling over and accounts for the speeding up of a spinning figure skater when she pulls in her arms.

Energy conservation prevents a swinging pendulum from going any higher without a push. When I taught introductory physics, I did a demonstration in which a large steel ball hung from a cable from the high ceiling of the lecture hall. I would stand with my head against the wall and hold the ball against my chin. Taking great care not to give it any push, I would release the ball and let it swing in a large arc across the room and then return to just miss my chin by a few millimeters. I trusted my chin to the law of conservation of energy. To hit my chin would have required it to have more energy than when it started. Actually, it lost a little energy to friction, so it did not quite return to where it started.

Newton's three laws of motion can be derived from momentum conservation alone, along with the definition of force as the time rate of change of momentum. Thus, almost all of Newtonian physics, or *classical physics*, is covered by the conservation laws. All that needs to be added to complete classical physics are the gravitational and the electromagnetic force laws. Fields that deal with many particles, like thermodynamics and fluid mechanics, are completely derivable from classical mechanics, and thus from the three conservation laws. Only with quantum mechanics do changes have to be made.

The conservation laws apply only to an isolated system that is not interacting with some other system. If the bodies that compose a system interact with other bodies, then their masses, linear and angular momenta, and energies can change with time.

So, when I define matter as stuff that kicks back when you kick it, I refer to the basic process by which we detect a material body. If you are walking in the dark and stub your toe on a rock, you have detected that rock by kicking it and experiencing it kick back. By kicking it you have imparted a momentum impulse to the rock and, because of momentum conservation, it gives that momentum impulse right back to you.

In daylight you can see the rock by it deflecting light from the sun into your eyes. Light is composed of material particles called *photons* that carry momentum, and that momentum changes because its direction changes when it is deflected by the rock.

This process of imparting momentum to bodies and measuring the returned momentum is repeated in all scientific instruments in one form or another. We not only can measure the momentum returned but also the energy returned, which in the case of a visible photon detected by our eyes specifies *color*.[11] From these in principle simple measurements we build up our picture of the material universe.

Since careful, quantitative experimentation began about four hundred years ago, we have not found any violations of the conservation laws. How might such a violation be observed? Suppose we were to simply see a body originally at rest suddenly, spontaneously begin to move without any visible input momentum. We would interpret it as a miracle. It might be attributed to the foot of God, such as kicking aside the Red Sea (or was it the Sea of Reeds?) so that the Israelites could escape Egypt. No reliably documented miracles have ever been reported in history or science.

In a later section I will show that these laws were not handed down by God but follow automatically and necessarily from the natural symmetries of space and time and the need for physicists to formulate their models in an objective way.

WHAT ABOUT SPIRIT?

If matter is something that kicks back when we kick it, then what might *spirit* be? Most people believe that something exists besides matter, stuff usually called spirit that is generally associated with mind and soul. All of chapter 8 will be devoted to the question of mind and whether it is possible for all our mental processes to be generated by matter alone with no soul or spirit. Here let us just ask what kind of stuff spirit might be that is separate from matter and yet interacting with it. Metaphorically, we might think of the "foot of God" giving us a kick once in a while. If God acts in the universe, such as by altering the path of a storm in answer to some petitioner's fervent prayer, then he gave that storm a kick.

Well, it may seem trivial, but I did not define matter as the stuff that gives us kicks. Something has to kick it first. Gods and ghosts don't kick back when you kick them. If they exist, they kick you whenever they want. (That's why Casper can walk through walls. The walls don't kick him the way they kick us when we try the same trick. Why doesn't he fall through the floor, then? He has no mass since he is no matter, and so has no weight.)

Now, I can put this on a much more precise footing (pun intended). As we saw above, a kick is composed of an impulse of momentum that can be quantified. Similarly, the retuned impulse, if no other bodies are involved, will be exactly the same magnitude in the opposite direction, conserving the momentum of the system.

In general, material bodies are objects that interact with other material bodies in a lawful way, conserving energy and momentum. Spirit is some kind of stuff that interacts with material bodies in a nonlawful way. It can violate our laws, which is another way of saying it can perform miracles. So there is the difference between matter and spirit: matter does not perform miracles; spirit does. If we ever saw a miracle, then that would be evidence for spirit. So far we have not.

THE UNIVERSE OF MATTER

The Hubble Space Telescope has detected galaxies as far as 13 billion light-years away. Since the light from that galaxy has taken 13 billion years to reach us, and the current estimated age of the universe is 13.7 billion years, then we are looking back to only 700 million years after the big bang. The luminous matter that we see at all distances from Earth, and at all times for the last 13 billion years, looks just like the matter we investigate in our laboratories on Earth. It is composed of the same stuff and obeys the same laws. At the fundamental level most of our familiar stuff is composed of just three elementary particles: the electron, the *u*-quark, and the *d*-quark. The quarks make up the nuclei of atoms, such as oxygen and iron (that is, the chemical elements); atoms are composed of nuclei surrounded by a cloud of electrons; atoms combine to make molecules, such as water and sugar.

With microscopes and other tools we can investigate the nature of the matter that composes living cells. That, too, is made of the same stuff—electrons and quarks—and it obeys the same laws. No special "vital force" is necessary to explain the properties of life. Life is a very complex aggregation of atoms that has evolved by natural selection many remarkable abilities such as reproduction, growth, the ability to absorb and use energy, self-healing, and, in at least a few species, most notably ours, thinking.

Our high-energy particle accelerators allow us to measure distances at the subatomic level, distances smaller than the size of the nuclei of atoms. In the 1960s and 1970s this capability provided physicists with a deep understanding of the elementary particles of matter and physical processes by which they interact. All the known elementary particles and the forces between them have been successfully described by what is called the *standard model of particles and forces.* The elementary particles of the standard model are shown in table 7.1. These interact by gravitational force, the electroweak force, and the strong nuclear force. The latter two forces are "mediated" by the exchange of the *bosons* shown in the table. The electroweak force is a unification of electromagnetism and the weak nuclear force, just as electromagnetism is a unification of the electric and the magnetic forces. Physicists hope someday to have a "theory of everything" (TOE) in which all the fundamental forces are unified as a single force. This is the situation they believe existed when our universe first appeared from chaos.

Although most familiar matter is composed of just the electron and *u*- and *d*-quarks, short-lived particles composed of the other elementary particles in the table appear at high energies, in accelerators and in cosmic rays. High-energy *muons* (basically heavy electrons, designated by μ in the table) are produced from the collisions of cosmic particles from outer space with the top of the atmosphere. One muon passes though every square centimeter of your body every minute. The most numerous particles in the universe are *photons* (γ in the table) and *neutrinos* (in three varieties signified by ν_e, ν_μ, and ν_τ in the table). These remnants of the big bang are a billion times more numerous in the universe than atoms. The photons form part of the 2.7-degree cosmic microwave background that was discovered in 1964.

As of this writing, the highest-energy particle accelerator yet built, the Large Hadron Collider (LHC), is going into operation in Geneva, Switzerland (with a part of the complex underground in France). Physicists are confident that the data from this machine will enable them to push to the next level of understanding the physical universe.

Fermions (antiparticles not shown)				Bosons
Quarks	u	c	t	γ
	d	s	b	g
Leptons	ν_e	ν_μ	ν_τ	Z
	e	μ	τ	W

Table. 7.1. The elementary particles of the standard model. Shown are the three generations of spin 1/2 fermions—quarks and leptons—that constitute normal matter and the spin 1 bosons that act as force carriers. Each has an antiparticle, not shown. The electroweak force is carried by four spin 1 bosons—γ, *W*⁺, *W*⁻, *Z*—and the strong force by eight spin 1 *gluons*, *g*. A yet unobserved spin zero *Higgs boson*, not shown, is also included in the standard model. (The general reader need not worry about these details; they are provided for completeness.)

THE MASS OF THE UNIVERSE

Although everything we see around us can be described in terms of atoms composed of quarks and electrons, luminous atomic matter constitutes only 0.5 percent of the mass of the universe. Another 3.5 percent of the mass of the universe is composed of nonluminous matter (like planets and dead stars) made of the same basic stuff. Most of the universe, however, is actually composed of two other ingredients that have not yet been identified: *dark matter* constitutes about 22 percent of the mass and *dark energy* predominates with about 73 percent. While these components remain unidentified at this writing, there is no doubt that they are material in nature since they produce gravitational "kicks," which is how they are indirectly observed.[12]

In only the last few decades has observational cosmology become a precise science. The dark energy mentioned above was not discovered until 1997. It is gravitationally *repulsive* and is responsible for an accelerating expansion of the universe. The average total energy density of the universe, including all the rest and kinetic energy of matter and all the potential energy that arises from the mutual gravity of matter, is precisely zero. Thus no energy from outside was required to make the universe. The law of conservation of energy was not violated. In fact, there is no need to invoke the violation of any physical law to account for the universe.

THE BEGINNING OF TIME?

Theists and theologians for years have persisted in making a fundamental physics error that was pointed out over twenty years ago. They assert that the start of the big bang was a "singularity" that marked the beginning of time. Here's what Dinesh D'Souza says in his critique of New Atheism: "In a stunning confirmation of the book of Genesis, modern scientists have discovered that the universe was created in a primordial explosion of energy and light. Not only did the universe have a beginning *in* space and time, but the origin of the universe was also a beginning *for* space and time."[13] D'Souza tells us the theological significance: "There was no time before the creation, Augustine wrote, because the creation of the universe involved the creation of time itself. Modern physics has confirmed Augustine and the ancient understanding of the Jews and Christians."[14] Of course the big bang is hardly a stunning confirmation of Genesis, where the story of creation bears no resemblance to the big bang and every divinely reported fact mentioned is wrong.

Nevertheless, similar statements can be found in almost every book written by theists and theistic theologians that talks about the origin of the universe. The recent book by Ravi Zacharias asserts, "Big Bang cosmology, along with Einstein's theory of relativity, implies that there is indeed an 'in the beginning.' All the data indicates a universe that is exploding from a point of infinite density."[15]

I heard the same assertion made while I was taking a walk and listening on my iPod to the fine course on the philosophy of religion by philosopher and believer Peter Kreeft.[16]

Although this idea has been discussed for years in the literature, theists generally refer to a mathematical proof by cosmologist Stephen Hawking and mathematician Roger Penrose published in 1970. They showed that Einstein's theory of general relativity implied that the universe we observe was once a "singularity," an infinitesimal point of infinite mass. Theologians then argued that space and time themselves must have begun at that point.

However, over twenty years ago Hawking and Penrose agreed that a singularity did not in fact occur. Their calculation was not wrong as far as it followed from the assumptions of general relativity. But those assumptions had not taken into account quantum mechanics. In his blockbuster best seller, *A Brief History of Time*, which came out in 1988, Hawking says: "There was in fact no singularity at the beginning of the universe."[17]

D'Souza has glanced at *Brief History*, mining quotations that seem to confirm his preconceived ideas. He quotes Hawking as saying, "There must have been a Big Bang singularity."[18] D'Souza has lifted it out of context and given it precisely the opposite meaning of what Hawking intended. Hawking was referring to the calculation he published with Penrose in 1970, and D'Souza cut off the quotation. Here it is in full: "The final result was a joint paper by Penrose and myself in 1970, which at last proved that there must have been a big bang singularity *provided only that general relativity is correct* and the universe contains as much matter as we observe. [Emphasis added]"[19] Hawking continues:

> So in the end our [Hawking and Penrose] work became generally accepted and nowadays nearly everyone assumes that the universe started with a big bang singularity. It is perhaps ironic that, having changed my mind, I am now trying to convince other physicists that there was in fact no singularity at the beginning of the universe—as we shall see later, it can disappear once quantum effects are taken into account.[20]

The main promulgator of the false notion that the big bang was the origin of time is the Christian apologist and philosopher William Lane Craig, who has been writing about cosmology and theology and debating the existence of God worldwide for decades. I debated Craig in Hawaii in 2003 and pointed out this error, which he has never acknowledged and continues to ignore. The argument can still be found on his Web site.[21]

A NATURAL SCENARIO FOR THE ORIGIN OF THE UNIVERSE

There is no reason why the universe had to have a beginning. It can just as easily stretch back in time without limit, so there never was a beginning, just as it can stretch forward in time without limit so there is never an end. A billion-billion years from now the universe may be lifeless and nothing but neutrinos, but a clock made of neutrinos would still keep ticking away.

My favorite scenario for the natural origin of the universe has the feature of an unlimited past and future. It is based on a 1983 proposal by James Hartle and Stephen Hawking.[22] I have worked out fully mathematically a simplified form of the model[23] and have published this in a book[24] and a philosophical paper.[25] I will just give a brief outline here.

Basically, in this scenario our particular universe appeared by a process called *quantum tunneling* from an earlier universe that, from our point of view, existed limitlessly in the past.[26] Quantum tunneling is a well-established phenomenon forbidden by the laws of Newtonian physics in which a body is able to penetrate through a barrier. The earlier universe, again from our point of view, has been contracting in a way opposite to what we experience with the big bang. However, the arrow of time in any universe is determined by the direction of increase in the total entropy, or disorder, of the universe. So time's arrow in our sister universe actually points in the reverse direction from ours and both universes can be viewed as arising by quantum tunneling from "nothing."

I must admit that this picture of the origin of the universe is not widely recognized. However, after being in print for several years no physicist, cosmologist, or philosopher has yet pointed out any errors. I do not claim that this is *in fact* how the universe arose. I merely present it as a scenario consistent with all of our knowledge by which the universe occurs naturally and thereby closes a gap where a theist might want to insert God.

WHAT ABOUT THE LAWS OF PHYSICS?

When theists hear me say, as I did above, that no laws of physics had to be violated to make the universe and thus there was no need for a miraculous, supernatural creation, they almost always respond: "Then where did the laws of physics come from?"

The answer of most physicists is: they are just there, brute facts about nature. However, we can do much better than that. Most people, including the majority of physicists and philosophers of science, view the laws of physics, or the laws of nature in general, as rules for the behavior of matter. For example, the conservation laws discussed above limit the movement of bodies.

I have argued that the laws of physics are not rules handed down by God or built into the structure of the universe. They are human inventions. They are not restrictions on the behavior of matter. They are restrictions on what physicists may do when they formulate mathematical models to describe their observations of matter.[27] The statements that the laws of physics make about our observations of matter are about the symmetries and objectivity of nature. When a law is broken, it is because some symmetry has been broken or some observation is being described from a unique or a subjective point of view.

I find it quite amazing that most scientists and philosophers have not realized this fact. It has been known, at least for the basic conservation laws, for ninety years.

In 1918 a German mathematician named Emmy Noether proved that the laws of conservation of linear momentum, angular momentum, and energy follow automatically and must be part of any mathematical theory that does not single out any particular position in space, direction in space, or moment in time, respectively.[28]

Now, from the time of Copernicus we have known that Earth was not the center of the universe. Since then we have learned that neither is the sun. In fact, no point in space can be identified as special—any different from any other point in space. We have also found that the universe is spherically symmetrical, that is, it singles out no particular direction in space. And, we have also found that no special moment in time can be specified.

While the beginning of the big bang would seem to be such a time, in fact that beginning could have been any other time. As I showed above, time did not begin with the big bang, despite the claim of theists and theologians.

In fact, space and time are also just human inventions. Time is defined by what is measured on a clock. The distance between two points in space is defined by how long it takes, measured on a clock, for light in a vacuum to travel between the two points. For example, suppose it takes one billionth of a second, one nanosecond, for light in a vacuum to go from one point to another. Then the distance is one "light-nanosecond," which is roughly a foot (0.3 meter).

When a physicist is formulating a model to describe observations of particle positions at various times under some set of conditions, he must ensure that the model does not contain any special moment in time. For example, he can't propose that a body falls from a height of 20 meters in 2 seconds on Mondays and 2.1 seconds on Fridays. When he writes down a model that does not depend on any absolute time, that model will *necessarily* contain a quantity called energy that is conserved. In the example, the potential energy of the body at a height of 20 meters is the same regardless of the day of the week, and that will guarantee (within the theoretical calculation) that the body takes 2.02 seconds whenever the experiment is done.

Now, notice that while space, time, and physical models are all human inventions, the results of observation are not arbitrary. In the above example, the time measured for a body to drop 20 meters is 2.02 seconds, agreeing with the model. If the model did not give this result, it would be falsified.

The universe did not *have* to be this way. It *might* have had a special moment in time, position in space, or direction in space. A creator could have done it any way he wanted, and in the process provided us with some evidence for his existence. The fact that none of these special quantities exists is exactly what you would expect if the universe was not created, that it came from nothing.

INVARIANCE

Noether's theorem was proved ninety years ago and nobody disputes it. Thus the most important laws of physics are simply what they are because the universe has certain space-time symmetries: you can move the universe any distance in your models, you can rotate it by any angle, and you can zero your clock at any time and you will get the same universe. The property an object or system has when it is unchanged under some operation is called *invariance.*

The theist will argue that God gave it those symmetries. But she can't argue that those symmetries, and the laws that follow by Noether's theorem, are *evidence* for God. God wasn't needed. A universe that came from nothing, or always existed if we stick to one arrow of time, such as the mirror universes in the scenario described above, would have just these symmetries and just these laws.

This cannot be controversial, only not understood. However, now I will carry this idea further and enter into areas that are likely to generate more legitimate argumentation.

Let us assume that we can represent an event as a point in a four-dimensional space and time, as suggested in 1908 by the German mathematician Hermann Minkowski. Assuming that the universe has no special direction in space-time, that is, space-time rotational invariance, we can derive all of Einstein's theory of special relativity, including $E = mc^2$. Again we have a basic law of physics not handed down from above, but the necessary consequence of a universe having no special orientation in space-time.

In my 2006 book, *The Comprehensible Cosmos*, I proposed these ideas and generalized space-time symmetries to the more abstract space that physicists use to build their theories. Invariance to rotation in that space was identified with what physicists call *gauge symmetry,* a mathematical symmetry that was the most important principle discovered in the twentieth century. I showed mathematically that much of physics, as we know it, follows with few other assumptions. I cannot go further into this here, but I propose that already with the great conservation laws we see that plausible natural answers may exist to the question "Where do the laws of physics come from?" They can very well have come from nothing.

NOTES

1. See my previous book, *Quantum Gods: Creation, Chaos, and the Search for Cosmic Consciousness* (Amherst, NY: Prometheus Books, 2009).

2. John F. Haught, *God and the New Atheism: A Critical Response to Dawkins, Harris, and Hitchens* (Louisville, KY: Westminster John Knox Press, 2008), pp. xiii-xiv.

3. Charles Tart, "Six Studies of Out-of-Body Experiences," *Journal of Near-Death Studies* 17, no. 2 (Winter 1998): 73–99.

4. Victor J. Stenger, *Physics and Psychics: The Search for a World beyond the Senses* (Amherst, NY: Prometheus Books, 1990), p. 111.

5. Charles T. Tart, "A Psychological Study of Out-of-Body Experiences in a Selected Subject," *Journal of the American Society for Psychical Research* 62 (1968): 3.

6. Leonard Zusne and Warren Jones, *Anomalistic Psychology: A Study of Extraordinary Phenomena of Behavior and Experience* (Hillside, NJ: Lawrence Erlbaum Associates, 1982).

7. Stenger, *Physics and Psychics,* pp. 178–79.

8. Michael Rea, *World without Design: The Ontological Consequences of Naturalism* (Oxford: Clarendon, 2002).

9. Richard C. Carrier, "Defending Naturalism as a Worldview: A Rebuttal to Michael Rea's *World without Design,*" Secular Web, http://www.infidels.org/library/modern/richard_carrier/rea.html#1.

10. The speed of light in a vacuum, denoted by c, was recognized by Einstein to be an arbitrary constant. For example, if you measure distance in light-years and time in years, $c = 1$ light-year per year.

11. Although the eye itself is sensitive to the one-photon level, several photons are generally needed for the brain to distinguish a signal from noise.

12. The dark energy may be associated with Einstein's cosmological constant, which corresponds to the curvature of empty space. However, it has a material equivalent and in any case still a purely natural phenomenon found in the equations of physics.

13. Dinesh D'Souza, *What's So Great about Christianity?* (Washington, DC: Regnery, 2007), p. 116.

14. Ibid., p. 123.

15. Ravi K. Zacharias, *The End of Reason: A Response to the New Atheists* (Grand Rapids, MI: Zondervan, 2008), p. 31.

16. Peter Kreeft, *Faith and Reason: The Philosophy of Religion,* Modern Scholar audiobook (Prince Frederick, MD: Recorded Books LLC, 2005).

17. Stephen W. Hawking, *A Brief History of Time: From the Big Bang to Black Holes* (New York: Bantam, 1988), p. 50.

18. D'Souza, *What's So Great about Christianity?* p. 121.

19. Hawking, *A Brief History of Time,* p. 50.

20. Ibid.

21. William Lane Craig, "The Existence of God and the Beginning of the

Universe," *Truth: A Journal of Modern Thought* 3 (1991): 85–96, http://www.leaderu.com/truth/3truth11.html (accessed November 22, 2008).

22. James B. Hartle and Stephen W. Hawking, "Wave Function of the Universe," *Physical Review* D28 (1983): 2960–75.

23. Following David Atkatz, "Quantum Cosmology for Pedestrians," *American Journal of Physics* 62 (1994): 619–27.

24. Victor J. Stenger, *The Comprehensible Cosmos: Where Do the Laws of Physics Come From?* (Amherst, NY: Prometheus Books, 2006), pp. 312–19.

25. Victor J. Stenger, "A Scenario for a Natural Origin of Our Universe," *Philo* 9, no. 2 (2006): 93–102, http://www.colorado.edu/philosophy/vstenger/Godless/Origin.pdf (accessed November 21, 2008).

26. I do not use the term "infinite" here since this is technically incorrect. The set of integers is an infinite set, but none of the members is "infinity."

27. The details are worked out fully mathematically in *The Comprehensible Cosmos.*

28. Nina Byers, "E. Noether's Discovery of the Deep Connection between Symmetries and Conservation Laws," Israel Mathematical Conference, http://www.physics.ucla.edu/~cwp/articles/noether.asg/noether.html (accessed February 20, 2009). This contains links to Noether's original paper, including an English translation.

8. THE NATURE OF MIND

So it is that the brain investigates the brain, theorizing about what brains do when they theorize, finding out what brains do when they find out, and being changed forever by the knowledge.

—Patricia Churchland[1]

THE IMMATERIAL SOUL

We have seen that science provides plausible scenarios for a purely material universe, with no need to introduce supernatural causes for any physical event, including the origin of the universe. Furthermore, the intuition that design is evident in the cosmos and in earthly life has been seen to have no basis and is no candidate for an irreducible gap that must be filled with God. The one place where theists still believe they see a gap in scientific knowledge into which they can try to insert God is the *mind.* They hope they can prove that the mind is not reducible to matter and therefore must have a supernatural component, traditionally known as the *soul.*

The notion that every human being has an immortal soul, or what is essentially equivalent, a *spirit*, is fundamental to almost all religious thought.[2] It was traditionally believed to be the essence of self, the source of one's personality, and the reservoir of all mental activity. All thinking was at least once thought to be done by the soul. All memories and emotions supposedly resided in the soul. All decisions, especially the exercise of free will, were performed by the soul. Thus it was the soul that was punished for freely committed sins. Very important, the soul was able to communicate directly with God or some cosmic consciousness. Furthermore, in many cultures, especially in Asia, the soul was and still is believed to be the force responsible for life—the *élan vital*, the source of vegetative as well as mental activities. The body dies when the soul leaves it because its life force is gone.

Now this is just a list of some of the traditional attributes of the soul. Not every culture and not every person in that culture held to every one of these beliefs in the past. Modern humans probably hold even fewer of these beliefs. But belief in a soul does entail at least some of these properties. We saw in chapter 4 how the Catholic Church grudgingly accepts the possibility that the body evolved from simpler matter but in no way extends that to the mind, which they equate at least in part with soul.

In many ancient cultures the soul was associated with breath. The Greeks called it *psyche*, which is related to *psychein*, meaning "to breathe." In Latin the soul is called *anima,* which can mean "air," "breath," or "life." The Latin word *spiritus* is also related to breathing and is the obvious source of "spirit." The Hebrew word for soul is *nephish*, which is also translated as "life" and connected to breathing. Also in Hebrew, *ruah* is translated as "wind" or "breath," and sometimes "spirit," "soul," "life," or "consciousness."

The association of soul or the life force with breath also occurred in the isolated Hawaiian Islands. According to one undocumented story from an oral culture, in old Hawaii when someone died native shamans tried to breathe life back into the body by shouting, "Ha!" in the body's face. When Western doctors came to Hawaii and were observed not to do this, they were called, "Ha-ole," which meant "without breath." To this day whites are called "haoles" in Hawaii.

So, for most of history the soul was considered to be part of the body—as material as the wind. The ancient atomists assumed the soul was made of atoms just like everything else. And Christianity maintained a unity of body and soul, with Jesus being fully resurrected in body. If only a disembodied soul survived death, then why the big deal about the empty tomb? Even today, those who look forward to the Second Coming (see chapter 2) see the future Kingdom of God on Earth as populated by the saved, including them of course, in fully reconstructed bodies. Presumably these will not be the bodies they had at death at age eighty, but the ones they had at eighteen, corrected for any imperfections such as obesity or protruding ears. And, of course, any brain cells that may have been destroyed by alcohol or drugs will be restored.

Thomas Aquinas (d. 1274) believed the soul was separate from matter, and this was one of the few of his teachings the Catholic Church did not adopt as dogma. The teachings of John Calvin (d. 1564) also implied an independent soul, while those of Martin Luther (d. 1546) did not. Calvin said souls would be with God immediately after death, which implied being separate from body. Luther said they slept until the Last Judgment.

Dualism was elaborated by the French philosopher René Descartes (d. 1650), who viewed the body as a machine and the mind, or soul, as a non-material entity that is not bound by the laws of physics. Philosophers now call Descartes' view *substance dualism.* Descartes speculated that the soul interacted with the body at the pineal gland, because of its central location in the brain. Neither he nor any dualist since has explained how the immaterial soul interacts with the material body except to simply give the usual theist fallback position, "God does it," as if that explains anything.

Baruch Spinoza (d. 1677) rejected substance dualism, arguing that there was only one infinite substance of which mind and body were different properties. Philosophers now call this *property dualism.* This is basically the teaching of the Catholic Church today, that a human being is "two in one, a divisible but a vital unity."[3] Sort of like the Trinity, only with two instead of three. At the same time, Catholics claim that the soul goes to heaven, hell, or purgatory immediately after death and does not get reunited with the body until the Last Day, with those in purgatory moving up to heaven at that time.

EVIDENCE OF ABSENCE

According to naturalism there is no immaterial soul, and mind is the product of matter and nothing more. All our thinking processes are the result of physical operations in the brain. Philosophers call this theory of the brain *physicalism.* What can be said about the physical nature of mind? A lot. If we go through every item on the list of attributes of the soul given above, we find it can be associated with physical processes in the brain. In the case of élan vital, the situation is even simpler; a special life force cannot be found anywhere in biology.

Drugs, brain injuries, aging, and other physical processes have a strong influence on thinking. If consciousness is spiritual, how does anesthesia render one unconscious? Thousands of experiments have been conducted in neuroscience laboratories where functional magnetic resonance imaging (fMRI) and other techniques are used to locate specific areas in the brain that become active for different kinds of thoughts, emotions, and decision making.[4] And before fMRIs, it had long been known that electrical stimulation of a point on a person's brain might elicit a memory or perception specific to that point and that person. This was discovered in the course of operations to cure severe cases of epilepsy. When these areas of the brain are damaged or surgically removed, those specific thoughts do not recur. Why should that be if we have an immaterial, thinking soul?

A very common phenomenon in humans is the *religious experience,* and it constitutes the main reason some people are totally convinced that God exists. They have had such experiences. The religious experience can happen during intense meditation or prayer, or during a trauma such as being near death. Various forms are also known as *psychic* experiences, *near-death* experiences, or *out-of-body* experiences. Those who have such encounters are often changed for life. They become true believers who "know" for sure that God and the hereafter exist. I discussed these experiences briefly in chapter 7 in the context of the validity of naturalism and will bring them up again in chapter 9.

Many theologians argue that religious experiences are evidence for the existence of God. Yet similar experiences can be triggered by electrical or magnetic pulses, or by loss of oxygen to parts of the brain, as during an epileptic fit.[5] When that part of the brain is surgically removed, the experiences no longer occur. If the soul exists, it seems to need the brain to talk to God. How is it going to talk to him after death? More likely, it's the brain doing all the talking to itself and that's where all the evidence points.

It is safe to say that there is absence of evidence for an immaterial, disembodied soul. Now, as I have frequently noted, I always hear, "absence of evidence is not evidence of absence." This is a cliché that has had its day. In my 2007 book, *God: The Failed Hypothesis*, I argued that there are many times when absence of evidence can be taken as evidence for absence. That occurs whenever the evidence that is absent *should* be there if your hypothesis is correct.

If the immortal, immaterial soul exists and is as important as people believe, then we should have evidence for it. Theologians assert that the religious experience itself is evidence for a soul. As we will now see, this is a hypothesis that is testable by scientific methods.

TESTING THE HYPOTHESIS

How might the validity of the religious experience be tested? Surprisingly, very easily. All that has to happen is that the person returning from such an experience report some fact that she could not have known ahead of time. This could be the successful prediction of some future event, like the exact date of the next Los Angeles earthquake.

Paranormal literature, which falls outside mainstream science, is filled with reports of people returning from near-death or out-of-body experiences with just such information. For example, you will read about someone on an operating table floating up above her body and able to look down to see what was going on and providing information, such as doctors' names read off their badges that she could not have known—especially since she was blind! Recall the Tart experiment discussed in chapter 7, where his subject was able to read a five-digit number on a shelf above her head. Or you will read about people reporting previous lives in "full detail," as in the famous Bridey Murphy hoax.[6]

But you have to read these reports with a large bag of salt. Every such report that has been examined by skeptics has a simple, natural explanation. Often it is simply a made-up story, such as the one about the floating blind woman.[7] In the Tart experiment, we found that the number was visible in its reflection from a wall clock. No carefully controlled scientific experiment that has ever been done to test these types of religious or psychic experiences, and there have been many, has found a confirmed positive effect. No one has ever come back from a religious experience with profound knowledge that could not have been in his head all along.

In the near-death experience the subjects all tend to see a tunnel with light at the end that they interpret as heaven. But you don't have to be near death to see this. Pilots being trained for high acceleration in a centrifuge report the same phenomenon, which is interpreted as the result of oxygen deprivation in the brain.[8] This is the same explanation that is given for the "visions" experienced by epileptics. Hippocrates called epilepsy the "divine disease." Perhaps some of the founders of the great religions were epileptics.

One of my most frustrating experiences in reading much of the literature on religion, including many serious scholarly works, is how many of these very capable authors take the reports of these kinds of "psychic" experiences seriously. For example, I recently listened to the audio of an excellent course on the philosophy of religion by philosopher Peter Kreeft of Boston College. Although a believer, he very evenly and fairly presents the philosophical arguments of both atheists and theists. In fact, I found his arguments for atheism much more convincing than those for theism, but no doubt a theist would think the opposite, and Kreeft would be happy with that because he forces both sides to listen to the other's arguments. Nevertheless, he repeatedly refers credulously to paranormal experiments, especially those done on near-death experiences. He also takes the reports on past-life regression seriously. I find it disheartening that a man of such clear intellect and willingness to examine all sides of arguments would not have had doubts raised in his own mind about reports that are clearly outside the mainstream of science and published in journals of minimal respectability. It illustrates how important a good science education is for any scholar.

PSYCHIC PSEUDOSCIENCE

I examined the whole field of psychic studies in my 1990 book, *Physics and Psychics*,[9] and brought that study up to date in 2003 with a chapter in *Has Science Found God?*[10] If the mind has a supernatural component, you would expect it to possess some special powers, to be able to overcome the laws of physics.[11] For over 150 years thousands of attempts have been made in scientific laboratories as well as in the field (such as haunted houses) to find evidence for psychic forces. Although many claims are published in books and specialized fringe journals sympathetic to the notion, the fact remains that these phenomena have not been accepted as confirmed by the consensus of the scientific community applying normal standards. In any other field of science, failure to find a phenomenon after such a long and diligent search would justify rejecting the existence of the phenomenon.

I get arguments from God believers about what evidence "should" exist if the soul exists. Why, they will ask, should the soul show those particular powers that have been tested in psychic studies? Well, maybe it shouldn't show these. But shouldn't the soul show something? As it stands, it shows nothing. The soul looks just as it should look if it does not exist.

At the time he wrote *The End of Faith*, Sam Harris thought that psychic phenomena were supported by the evidence: "There also seems to be a body of data attesting to the reality of psychic phenomena, much of which has been ignored by mainstream science."[12] He refers specifically to a 1997 book by Dean Radin called *The Conscious Universe: The Scientific Truth of Psychic Phenomena*.[13] I discussed Radin's claim in *Has Science Found God?*[14]

Radin does not base his assertion on any individual experiments that meet all the necessary requirements for a major new discovery in any legitimate scientific field: statistical significance, all systematic errors ruled out, independent replication, and others. Not a single psychic experiment in 150 years has managed this.

Rather, Radin used a dubious method called *metanalysis* in which the results from many experiments, good ones *and bad ones*, are combined and analyzed as a whole. Although each by itself is insignificant, the net result is supposed to be significant.

Experts in statistics uncovered a number of errors in Radin's analysis, rendering his conclusions useless.[15] While metanalysis is used to seek out trends in social sciences, I have never heard of a single major discovery made with metanalysis. After forty years of active scientific research on fundamental questions, I am sure I would have known if even one had been.

Sam Harris has since admitted that he lacks the knowledge to assess the validity of psychic claims. As we saw above, he has clarified his position on mysticism and spiritualism. He has added this on metaphysical questions:

> There is simply no question that people have transformative experiences as a result of engaging contemplative disciplines like meditation, and there is no question that these experiences shed some light on the nature of the human mind (any experience does, for that matter). What is highly questionable are the metaphysical claims that people tend to make on the

> basis of such experiences. I do not make any such claims. Nor do I support the metaphysical claims of others.[16]

THOUGHTS AND MATTER

I have already sufficiently quoted theologian John Haught to demonstrate that he does not support naturalism or physicalism. He calls naturalism "deeply self-contradictory."[17] He does not, however, provide any specific contradiction. The best Haught can do is assert his personal judgment that evolution, in particular, will never be able to explain certain mental phenomena such as cognition. He claims, "Scientific naturalism ignores the subjective side of nature, especially our inner experience."[18]

I do not think it is fair to say that scientific naturalism ignores the subjective. While it is true that neuroscientists do not yet have an established material model of mind, they have considerable data on changes that occur in the brain during subjective mental activity. They establish beyond doubt that material processes are involved.

This seems to be the common refrain of theists arguing against a purely natural universe. In his book *God Is No Delusion: A Refutation of Richard Dawkins*, Thomas Crean, a Dominican friar of the priory of St. Michael the Archangel in Cambridge, follows Haught in using the argument from ignorance, saying he cannot understand how thoughts could emerge from matter. He asks, "How could a 'material kind of thing' cause an 'immaterial kind of thing' to exist?"[19] Well, a computer is a material kind of thing that can solve mathematical and logical problems. It can write poetry that English professors are unable to distinguish from that written by humans. It can produce beautiful art and music. The aesthetic experiences of these are immaterial kinds of things, but they result from physical brain activity.

Unaware of these facts, Crean continues in that vein, "Materialism, then, is absurd. A thought cannot be a material thing, nor can it be caused by a material thing. The only possible conclusion is that thought as such is something independent of matter, that is, something *spiritual*."[20] These are just illustrations of Crean's own ignorance.

In a recent short book misleadingly titled *Naturalism*, philosophers Stewart Goetz and Charles Taliaferro attempt to show that naturalism is intellectually incoherent.[21] The title is misleading because the book is not about naturalism so much as it is an apologetic for Christian theology.[22] The authors are theists who teach at Notre Dame University and St. Olaf College, respectively. They make the claim that a duality of the physical and the mental is necessary to explain mental causation, that is, how mental events cause physical events.

This strikes me as rather backward. If, as naturalism teaches, mental events arise from physical events in the brain, then there surely can be no problem since we then have physical events causing physical events, just as a cue ball can hit an eight ball and cause it to go into a pocket. On the other hand, if mental events have their own nonphysical nature, then we have the problem of explaining how they cause physical events. Goetz and Taliaferro do not provide us with even a speculative model for how that can happen.

When theists such as Goetz and Taliaferro refer to gaps in the scientific record, the best they can do is say, "See, God must have done it." This provides no more information and is less economical than a simple statement: "Nature did it." But atheists can usually do much more than this simple assertion and give some idea of how nature did it. In a physical explanation we often have a theory such as relativity or evolution that provides detailed mechanisms for the events being observed. Even where we do not have an existing established theory, such as for the origin of life or mental processes, we have plausible proposals under consideration that agree with all existing knowledge and that require no supernatural elements. Theists can make only the simple assertion, "God did it." Scientists can say: "We don't know. But we'll try to find out."

Of course, mind-body dualism is a widespread "commonsense" belief among laypeople. You know what common sense is, don't you? It's the human faculty that tells us the world is flat. Goetz and Taliaferro seem to think common sense is sufficient to adopt the dualist view, explaining in turgid fashion:

> If a person is convinced that his reasons for believing that he is a nonspatial entity and that he causally interacts with a physical body are better than any reasons he is given for believing that there can be no noncausal pairing relations between a nonspatial soul and a physical body that makes possible causal interactions between the two, then he will be

justified in asserting the existence of such a relation, *even though he does not know what it is.*[23]

In other words, a person can believe whatever he wants to believe even if he doesn't know what it is he believes.

Goetz and Taliaferro also claim to show the philosophical coherence of divine agency. So what if it is philosophically coherent? That says nothing about its reality. A fantasy computer game in which heroes come back to life after being killed is philosophically coherent. It wouldn't run on a computer if it wasn't logical. But the world is still not that way.

Philosopher Paul Churchland points out that all throughout history people have expressed doubt that science would ever be able to explain some phenomena.[24] The first-century astronomer Ptolemy (c. 85–165), the greatest astronomer of the age, said science would never be able to capture the true nature of heavenly causes because they were inaccessible. He didn't have Newton's inspiration that the laws of physics are universal, applying both on Earth and in the heavens. The nineteenth-century philosopher Comte (d. 1857) similarly argued we could never know the physical constitution of stars. He didn't know about atomic spectra. As late as the 1950s most people were still expressing doubt that life could be explained purely materialistically but that some life force was needed. With the discovery in 1953 of the structure of DNA and the great success of the theory of evolution by natural selection, science saw no need for, and indeed no evidence for, a special force of life.

When the mental dualist asks, "How can thoughtless matter give rise to thought?" he is expressing the same argument from ignorance used by those who say, "How can dead matter give rise to life?"

Now, I should note that not everything that we talk about in our materialistic theories is itself composed of matter. Any theory uses abstract ideas that are not made of matter. An equation is immaterial; so is a distance or a number. That material objects enter into relations and relations are immaterial does not imply that relations can exist independent of material objects. If fact, they don't. Take the matter away and you don't have gravitational attraction or conservation of energy. Nor do you have thought.

Many questions remain unanswered by those who claim that some immaterial spirit or soul is ultimately controlling the actions of the brain.

How does this immaterial thing that carries no energy or momentum provide energy and momentum to particles in the brain? This implies violations of conservation of energy and momentum, which the theist believes are God's laws. Why is it okay to break these laws of God and not his other laws, such as no homosexual marriage or no using condoms? But if spirits carry energy and momentum, then they are material entities.

PHILOSOPHICAL MUSINGS ON MIND

Philosophers of mind are far more active in interpreting the work of the science of mind than they are in the more highly developed scientific fields of physics and biology. According to Paul Churchland, the main problem is our lack of understanding of what constitutes *consciousness*.[25] We can readily accept that various complex electro-chemical-mechanical processes take place in our bodies. For example, the material nervous system of the body operating in concert with various organs controls our hands and feet. While we may consciously decide to move a hand or a foot, we certainly do not have to worry about all the detailed physics and chemistry that is taking place and calculate the force needed. This is all done unconsciously. These phenomena can be viewed *objectively*, just like those of any machine. That is, they can be studied by various observers from different points of view.

However, consciousness is *subjective*, available to only the one person who is experiencing it. And we have, in Churchland's words, no "clear and evocative model that promises some useful grasp of its essential nature."[26]

Going back to Gottfried Leibniz (d. 1716), philosophers have made attempts to demonstrate the problem by means of narratives or "thought experiments." I will use the narrative known as the *knowledge problem*, published in 1982 by the Australian philosopher Frank Jackson.[27]

A neuroscientist named Mary has been raised artificially since birth to see only in black and white. However, in her training she has learned all there is to know about the physical nature of color and how the human visual system and brain handles it. Yet her science does not tell her anything about the conscious experience of color. When she is released from her restriction to black and white and looks at a ripe tomato, she experiences redness for the first time. Jackson concluded that there must be something unphysical about that experience since it was not included in her physical knowledge of color and how the body handles color data.

New atheist Daniel C. Dennett responds that if Mary really knew everything there was to know about the physical nature of color, she would be effectively omniscient and be able to deduce her own reaction to it.[28] Thus it is purely physical. I agree. However, it seems to me that Jackson has no right to introduce a thought experiment that gives his character unrealistic powers. Why not just put God in there instead of Mary? Then his conclusion that the process is nonphysical is true by definition.

In a more reasonable narrative, Mary does not know everything and cannot reconstruct the experience, so when she is released from her constraint and observes the tomato, she has the experience of redness for the first time. Since she is not omniscient, she is still open to new knowledge and there is no reason why that knowledge cannot still be physical, why it cannot still result from a material process in the brain that had not been activated earlier.

It would be like my experience as a kid playing baseball. I thought all the time about what it would feel like to hit a home run. When I finally did hit one (without steroids), I experienced it for the first time.

The experience of color is an example of what philosophers call *qualia*, which are qualities of experience, such as redness or pain. As philosopher John Searle describes the experience of pain,

> Suppose we tried to say that pain is really "nothing but" the patterns of neuron firings. Well, if we tried such an ontological reduction, the essential features of pain would be left out. No descriptions of the third-person, objective, physiological facts would convey the subjective, first-person character of the pain, simply because the first-person features are different from the third-person features.[29]

Churchland responds that Searle is simply assuming what he sets out to prove by asserting that first-person features are different from third-person ones. And even if they are, why can't the first-person sensation be physical in nature?

LOOKING AT THE DATA

Psychologists and neuroscientists are beginning to come up with plausible answers to the fundamental question that has been raised by theologians and philosophers: how can matter create thought? A recent book by neuropsychologist Chris Frith of the University of London called *Making Up the Mind* takes a big step in that direction.[30] Here is how Firth summarizes his results, which are not based on philosophical musings but on an examination of the data we now have about the brain and how it operates:

> In this book I shall show that this distinction between the mental and the physical is false. It is an illusion created by the brain. Everything we know, whether about the physical or the mental world, comes to us through our brain. But our brain's connection with the physical world of objects is no more direct than our brain's connection with the mental world of ideas. By hiding from us all the unconscious inferences that it makes, our brain creates the illusion that we have direct contact with objects in the physical world. And at the same time our brain creates the illusion that our own mental world is isolated and private. Through these two illusions we experience ourselves as agents, acting independently upon the world. But at the same time, we can share our experiences of the world. Over the millennia this ability to share experience has created human culture that has, in its turn, modified the functioning of the human brain.
>
> By seeing through these illusions created by our brain, we can begin to develop a science that explains how the brain creates the mind.[31]

Frith's claim is supported by a growing body of experiments in which a subject's conscious awareness presents a false picture of reality. Some of these are already familiar, such as subliminal perception, blind-sight, and phantom limbs. Perhaps the most startling example, one that clashes with our most cherished notions of self, is the evidence that the brain makes decisions before our conscious awareness of deciding to do something. Benjamin Libet and collaborators first showed this in a classic experiment in 1983.[32]

In the Libet experiment, which has since been independently confirmed,[33] a subject was simply asked to lift a finger whenever she felt inclined to do so. She was hooked up to equipment that measured electrical activity in the brain. She was asked to signal the moment she had the urge to lift her finger. That urge occurred about two-tenths of a second before the finger was actually lifted, showing an unsurprising delay between decision and action. However, brain activity indicating that the subject was about to lift her finger occurred five-tenths of a second before the finger was lifted, or three-tenths of a second *before* she reported she was consciously aware of deciding to lift the finger.

At first glance, this would seem to provide evidence against the existence of free will. Actually, it does not. It simply indicates that if we have free will, it is exercised by our unconscious rather than conscious mind. This has the theological implication that the soul, if it exists and is the source of our free will, must be associated with the unconscious mind and not just the conscious mind as has been usually assumed. If our unconscious mind sins, then shouldn't it be punished? The more we learn, the more incoherent religious beliefs become.

But we have no reason to make a connection between the unconscious and the soul. The main examples in Frith and thousands of other books and articles on empirical neuroscience provide convincing evidence that matter makes the mind. In one recent example of many, a kind of mind reading has been performed using external, objective, and physical fMRI data to reconstruct a subjectively reported image. This is significant because, as we saw above, one of the arguments made by those who think the mind is more than a material phenomenon is to question how a subjective experience can be described by the objective techniques of science.

NEUROTHEOLOGY

I predict that in the not too distant future we will have a purely material model of the brain that is consistent with all our observations, from daily life to the laboratory, and that leaves no room for any supernatural element. Rest assured it will be fought tooth and nail by the true believers who are unwilling to accept the verdicts of science and reason. We can soon expect a Michael Behe or a William Dembski of "neurotheology" to emerge, operating with generous funding from wealthy organizations such as the Discovery Institute and the Templeton Foundation. They will publish books from Christian publishing houses and speak to large, receptive audiences of the faithful around the country. They will attempt to get their "science" taught in science classes. Politicians will pass laws mandating "equal time" for their "alternative" theories that will be fought in court. They will lose this battle, just as they did with those fought in the guise of creation science and intelligent design.

In fact, as I write, this movement is beginning to take place.[34] On the Discovery Institute Evolution blog neurosurgeon Michael Egnor repeats the theological argument we heard above that ideas cannot be created from matter: "Clearly, under ordinary circumstances the brain is necessary for our ideas to exist, but, because matter and ideas share no properties, it's hard to see how the brain is *sufficient* for ideas to exist."[35] Of course this is, once again, the argument from ignorance. It's hard for Egnor to see how the brain is *sufficient* for ideas to exist, therefore the material brain is insufficient.

As we have seen, purely material computers are perfectly capable of producing ideas. Pressed by new atheist P. Z. Myers to be scientific rather than philosophical,[36] Egnor suggests a scientific test of his hypothesis in which the brain is subdivided into parts and the parts kept separately alive. He hypothesizes that each part would still retain the same ideas, such as altruism, as the whole.

While this exact experiment has obviously not been done, the equivalent is in fact very common. A wealth of data exists in which parts of the brain are injured or removed. If Egnor's hypothesis is correct, then those brain changes should not affect thought patterns. In fact, they do—dramatically. Egnor's hypothesis was falsified even before he proposed it.

Psychologist Mario Beauregard also relies heavily on the argument from ignorance to claim that gaps in neuroscientific knowledge he has uncovered in a number of peer-reviewed scientific publications can only be explained by duality.[37] He does not make the claim in the scientific papers themselves (where they would have been rejected) but in philosophical publications. There he states his personal opinion, without proof, that materialistic explanations are inadequate. He also claims that quantum physics plays a role and, forgetting that quantum physics is purely materialistic, falsely concludes that this requires a nonmaterial mind.[38] We will discuss quantum mechanics and the brain later in this chapter.

NEURAL NETWORKS

As we have seen, the brain clearly carries on certain mechanical functions that can be done on a machine. In fact, robots perform many of these functions. The issue is whether all the properties we associate with mentality can be done by a purely material system.

Indeed, the brain in many ways looks like a computer. Now, it does not at all resemble the familiar digital computer, which is based on an evolved form of the architecture originally proposed around 1945 by the great mathematician John von Neumann. However, this is just one of the possible architectures upon which one can build a computer. A computer can also be built that crudely models the brain, with a network of connected "neurons" that provide an output based on their inputs.[39] Simple artificial neurons might be made from the logic gates familiar in binary electronics. They each have one or two inputs and one output. Let A and B be the two inputs to a gate, each of which can have one of two values, the binary numbers 0 or 1 (or False or True). The output C is also 0 or 1. Here are six important examples:

OR: C = A or B. C = 1 if either A or B is 1, or if both A and B are 1.
AND: C = A and B. C = 1 if and only if both A and B are 1.
NOT: C = not A. C = 1 if A is 0, C = 0 if A is 1. In this case there is no B input.
XOR: C = A xor B. C = 1 if A or B is 1 but 0 if both are 1.
NOR: C = A nor B. C = 0 unless A and B are both 0.
NAND: C = A nand B. C = 1 unless A and B are both 1.

These gates plus a few other simple operations can be used to implement *Boolean algebra*, or *Boolean logic*, a logical calculus developed by George Boole in 1854. Any conventional logical expression can be expressed in Boolean terms.[40]

Actually, it can be shown that all logical operations can be performed with nor gates alone or nand gates alone.

This means that a neural network of simple logic gates is able to perform logical deduction, including most of mathematics with an exception discussed below. The neurons of the brain are far more complicated than simple logic gates, but they can readily perform that function. For our purposes, this proves that the material brain is at least capable of logical deduction without the aid of a soul.

While simple neural networks have been built, most neural network studies, which have been devoted to finding practical uses such as pattern recognition, are simulated on powerful digital computers. None have attempted to come close to duplicating even a small part of the brain. I am somewhat surprised, as an outsider, that no one has, to my knowledge, attempted to duplicate the brain of an insect, which I would guess is in the realm of technical feasibility.

Let me attempt to draw a comparison between the brain and an artificial neural network that might be built with current technologies. The human brain contains about 100 billion neurons. I'm only making a physicist's guess now, but I would imagine that the best we could do with current technology would be to build a network with about 100,000 artificial neurons, a million times fewer than the human brain. So we can't come close—yet—to the brain in neural capacity (although the Internet may have a billion nodes).

But the artificial machine has one big advantage over the brain, which accounts for that fact that it can do many things, such as serial computing, far faster than the brain. Signals between artificial gates are carried by electromagnetic pulses that move at the speed of light, 3×10^8 meters per second. In the brain, signals move by a complicated chemical process involving sodium and calcium ions no faster than about 130 meters per second. While modern computers perform operations in a billionth of a second or less, the brain operates on a timescale of about a thousandth of a second—a million times slower, with most reactions taking tenths of a second.

So, our artificial neural network would have a million times fewer neurons than the brain but be a million times faster. Clearly it is the large number of neurons of the human brain that enable it to do so much better than artificial machines, of any architecture, in certain tasks such as pattern recognition. Basically the brain does better in parallel processing, while the von Neumann machine does better in serial processing.

We can look at the brain with its 100 billion neurons and be impressed by its wondrous "design." But, we can also look at its incredible slowness and silly chemical communication system, and see in those facts yet more evidence for lack of design, intelligent or otherwise, in biology. In its usual

Rube Goldberg way, evolution cobbled together something that worked well enough and left it at that. In engineering, the term for a device put together in this haphazard fashion is called a "kluge." Psychologist Gary Marcus shows how the brain is a kluge in his recent book called, you guessed it, *Kluge*.[41] You would think an intelligent designer would have been capable of designing a brain using electrical currents. Dumb humans have put electricity to use for over a century now.

Maybe someday we will come in contact with an alien race that did evolve brains with the same number of neurons as we have but that signal one another at the speed of light. Imagine the power of such brains!

IS THE BRAIN A COMPUTER?

I have made the case that the purely material neural network of the human brain can easily perform logical and mathematical tasks. This includes the execution of most, if not all, computer algorithms. It remains to be seen if these tasks can encompass the various problems of consciousness and purposeful action discussed earlier, but we have demonstrated that none of these requires a god of the gaps.

In lengthy tomes published in 1989 and 1994, mathematician Roger Penrose made a deeply thoughtful attempt to show that the human brain is not solely a computer.[42] He based his argument on the fact that the brain performs noncomputational (nonalgorithmic) acts that cannot be executed on a computer. He used the example of mathematics where, according to a theorem derived by mathematician Kurt Gödel in 1931, unprovable truths can exist within any consistent formal mathematical system at least as complicated as arithmetic.[43] Mathematicians have been able to arrive at these truths by some means other than the conventional method of mathematical proof.

It has been shown that mathematical proofs can be recast as algorithms that can be performed on a computer. According to Penrose, it follows that a computer cannot do everything the brain can do. Computers perform algorithms. Mathematicians can think nonalgorithmically.

Penrose did not conclude that something supernatural was going on in the brain. Rather, he suggested that quantum mechanics, in particular quantum gravity, might play a role. However, as I showed in my 1995 book,

The Unconscious Quantum, an application of textbook quantum mechanics to the brain shows that it is too hot and its parts are too big for quantum effects to be significant. Physicist Max Tegmark arrived at the same conclusion in a 1999 paper.[44]

Penrose's claims have failed to gain the consensus of the philosophers of mind and scholars in the field of artificial intelligence.[45] I will focus on just one question: can a computer perform nonalgorithmic functions? Penrose did not really prove they could not, just that not all thinking is algorithmic. He wrongly assumed that all computers can do are algorithms.

Physicist Taner Edis has shown that randomness can be used to generate novelty and produce outcomes that are not predetermined by a set of rules.[46] This provides theoretical support for the suggestion I made in 1995[47] and expanded on in 2003[48] that random processes in the brain could be responsible for the appearance of purposeful mental actions.

As I mentioned above, the brain is not a quantum device. It operates for the most part according to Newtonian mechanics. The fact that the atoms in the brain operate on quantum principles does not change this conclusion. So do the atoms of a rock, and we are not considering rock consciousness. Most of the time the brain is "just" a computer carrying out deterministic mechanical algorithms in which a given input should always result in the same output.

Now, familiar events on the macroscopic scale obey Newtonian mechanics in which the motion of a body is completely predetermined by its initial conditions and the forces acting on it. Still, they are often observed to give different outcomes for what seems to be the same inputs. This is because it is often very difficult to precisely specify the initial conditions or repeat them exactly in every run of the experiment. Even the best bowler cannot make a strike each time.

Furthermore, complex physical systems are sometimes characterized by *chaos*, the phenomenon in which the tiniest change in initial conditions can result in dramatically different system behavior. This is called the "butterfly effect," where the flap of a butterfly's wings, metaphorically speaking, can change the weather days ahead. The Earth's atmosphere is an example of a chaotic system. The brain may exist on the "edge of chaos," where it is stable and predictable most of the time but capable of changing state rapidly under the right conditions.

All of these effects are possible in the brain, which is macroscopic in the sense that it is large, complex, and contains many particles. Because it is hot, thermal motion is considerable and this can cause the types of randomness that Edis talks about. I have suggested that radioactivity in blood could also produce random changes in the chemistry of cells. Blood is like seawater and contains radioactive potassium-40, an isotope that emits electrons with energies sufficiently high to break atomic and molecular bonds. Also, we have seen that a high-energy cosmic-ray muon produced in the upper atmosphere passes through every square centimeter of our bodies every minute. This could produce similar effects.

So the brain can easily be a nonquantum device, operating according to Newtonian physics and carrying out calculations and logical deductions but still functioning close to chaos with an ingredient of randomness caused by thermal motion or radioactivity. An occasional glitch can then redirect an algorithm in a different direction. This may be the source of creativity, even free will, although it is certainly not what most people think is happening in these situations. Rather, they imagine some kind of disembodied self as making all these decisions. The results are the same. How would you empirically distinguish between an apparently free choice and one that was made on the toss of dice? Ask the person? We have seen that the brain makes most of its decisions subconsciously, so the testimony for some claiming free choice could simply be the story made up by the brain.

SO, WHAT ABOUT THE SOUL?

There can be no more important question for human beings than whether we have souls. The answer tells us whether or not we will have life after death. I certainly do not claim to have settled the matter in the short discussion above. It sure looks like the brain can do the full job of mentality, but we are not near having the kind of well-established science we can point to as with the standard model in physics, our knowledge of cosmology, and Darwinian evolution in biology.

In these domains, we have fully developed and tested models in which the ingredients are matter alone. In the case of the soul, we have no direct evidence for its existence and no need to include it as either the vital force of life or as some integral component of the mind.

As mentioned, theist organizations such as the Discovery Institute and the Templeton Foundation are beginning to organize conferences and fund research into what they perceive as the remaining gap for God in scientific knowledge—the ultimate source of consciousness. With the rate of progress being made in neuroscience and the wonderful tools at their disposal, which keep getting better and better, such as functional magnetic resonance imaging, I am looking forward to living long enough to see that final gap closed by matter alone.

NOTES

1. Patricia Smith Churchland, *Neurophilosophy: Toward a Unified Science of the Mind-Brain* (Cambridge, MA: MIT Press, 1995), p. 482. First published in 1986.

2. For a nice summary of the history of the idea of soul, see Jerome Elbert, *Are Souls Real?* (Amherst, NY: Prometheus Books, 2000).

3. "Soul," *New Advent Catholic Encyclopedia*, http://www.newadvent.org/cathen/14153a.htm (accessed December 6, 2008).

4. My physicist son, Dr. Victor Andrew Stenger, is currently involved in such work at the University of Hawaii Medical Center.

5. Michael Shermer, *Why People Believe Weird Things: Pseudoscience, Superstition, and Other Confusions of Our Time* (New York: W. H. Freeman, 1997); Michael Shermer, "Why People Believe in God: An Empirical Study on a Deep Question," *Humanist* 59, no. 6 (1999): 43–50.

6. Melvin Harris, "Are 'Past-Life' Regressions Evidence for Reincarnation?" *Free Inquiry* 6 (1986): 18.

7. See my discussion and references in Victor J. Stenger, *Has Science Found God? The Latest Results in the Search for Purpose in the Universe* (Amherst, NY: Prometheus Books, 2003), p. 297.

8. Susan J. Blackmore, *Dying to Live: Near-Death Experiences* (Amherst, NY: Prometheus Books, 1993); Keith Augustine, "Hallucinatory Near Death Experiences," Secular Web, http://www.infidels.org/library/modern/keith_augustine/HNDEs.html#experiments (accessed December 11, 2008); G. M. Woerlee, *Mortal Minds: The Biology of Near-Death Experiences* (Amherst, NY: Prometheus Books, 2005); G. M. Woerlee, "Darkness, Tunnels, and Light," *Skeptical Inquirer* 28, no. 3 (2004), http://www.csicop.org/si/2004-05/near-death-experience.html (accessed January 11, 2009).

9. Victor J. Stenger, *Physics and Psychics: The Search for a World beyond the Senses* (Amherst, NY: Prometheus Books, 1995).

10. Stenger, *Has Science Found God?* ch. 10.

11. Parapsychologists insist that psychic phenomena need not be supernatural. We can argue about that when and if they ever find any.

12. Sam Harris, *The End of Faith: Religion, Terror, and the Future of Reason* (New York: Norton, 2004), p. 41.

13. Dean I. Radin, *The Conscious Universe: The Scientific Truth of Psychic Phenomena* (New York: HarperEdge, 1997).

14. Stenger, *Has Science Found God?*

15. I. J. Good, "Where Has the Billion Trillion Gone?" *Nature* 389, no. 6653 (1997): 806–807; Douglas M. Stokes, *The Nature of Mind: Parapsychology and the Role of Consciousness in the Physical World* (Jefferson, NC, and London: McFarland, 1997).

16. Sam Harris, "Response to Controversy," http://www.samharris.org/site/full_text/response-to-controversy2/ (accessed February 3, 2009).

17. John F. Haught, *God and the New Atheism: A Critical Response to Dawkins, Harris, and Hitchens* (Louisville, KY: Westminster John Knox Press, 2008), p. 82.

18. Ibid.

19. Thomas Crean, *God Is No Delusion: A Refutation of Richard Dawkins* (San Francisco: Ignatius Press, 2007), p. 27.

20. Ibid., p. 28.

21. Stewart Goetz and Charles Taliaferro, *Naturalism* (Grand Rapids, MI: William B. Erdman).

22. See review by Paul Draper, "Naturalism," *Notre Dame Philosophical Reviews*, http://ndpr.nd.edu/review.cfm?id=14725 (accessed November 18, 2008).

23. Goetz and Taliaferro, *Naturalism*, p. 64.

24. Paul M. Churchland, *The Engine of Reason, the Seat of the Soul: A Philosophical Journey into the Brain* (Cambridge, MA: MIT Press, 1995), pp. 187–90.

25. Ibid., p. 188.

26. Ibid.

27. Frank Jackson, "Epiphenomenal Qualia," *Philosophical Quarterly* 32 (1982): 127–36.

28. Daniel C. Dennett, *Consciousness Explained* (Boston: Little, Brown, 1991).

29. John R. Searle, *The Rediscovery of the Mind* (Cambridge, MA: MIT Press, 1994), p. 117. First published in 1992.

30. Chris Frith, *Making Up the Mind: How the Brain Creates Our Mental World* (Oxford: Blackwell, 2007).

31. Ibid., p. 17.

32. B. Libet et al., "Time of Conscious Intention to Act in Relation to Onset of Cerebral Activity (Readiness-Potential): The Unconscious Initiation of a Freely Voluntary Act," *Brain* 196, no. 3 (1983): 623–42; Frith, *Making Up the Mind*, pp. 66–68.

33. P. Haggard et al., "On the Perceived Time of Voluntary Actions," *British Journal of Psychology* 90, no. 2 (1999): 291–303.

34. See "Non-materialist neuroscience," Rational Wikipedia, http://rationalwiki.com/wiki/index.php?title=Non-materialist_neuroscience (accessed January 12, 2009).

35. Michael Egnor, "Ideas, Matter, and Faith," *Evolution News and Views*, http://www.evolutionnews.org/2007/06/ideas_matter_and_dogma.html (accessed January 12, 2009).

36. P. Z. Myers, "Egnor's Machine Is Uninhabited by Any Ghost," *Pharyngula*, http://scienceblogs.com/pharyngula/2007/06/egnors_machine_is_uninhabited.php (accessed July 12, 2009).

37. M. Beauregard and V. Paquette, "Neural Correlated of a Mystical Experience in Carmelite Nuns," *Neuroscience Letters* 405 (2006): 186–90; M. Beauregard et al., "Dysfunction in the Neural Circuitry of Emotional Self-Regulation in Major Depressive Disorder," *Neuroreport* 17 (2006): 843–46; J. Levesque et al., "Effect of Neurofeedback Training on the Neural Substrates of Selective Attention in Children with Attention-Deficit/Hyperactivity Disorder: A Functional Magnetic Resonance Imaging Study," *Neuroscience Letters* 394 (2006): 216–21.

38. J. M. Schwartz et al., "Quantum Physics in Neuroscience and Psychology: A Neurophysical Model of Mind-Brain Interaction," *Philosophical Transactions of the Royal Society B: Biological Sciences* (2005): 1–19.

39. For a good discussion, see Stan Franklin, *Artificial Minds* (Cambridge, MA: MIT Press, 1995), ch. 6.

40. Here I refer to the standard syllogistic logic developed by Aristotle and still commonly used. Other logics are possible but need not concern us here.

41. Gary F. Marcus, *Kluge: The Haphazard Construction of the Human Mind* (Boston: Houghton Mifflin, 2008).

42. Roger Penrose, *The Emperor's New Mind: Concerning Computers, Minds, and the Laws of Physics* (Oxford: Oxford University Press, 1989); *Shadows of the Mind: A Search for the Missing Science of Consciousness* (Oxford: Oxford University Press, 1994).

43. Kurt Gödel, "On Formally Undecidable Propositions of *Principia Mathematica* and Related Systems (title translated)," *Monatshefte für Mathematik und Physik* 38 (1931): 173–98. English translation in Jean van Hejenport, ed., *A Source Book in Mathematical Logic, 1879–1931* (1967), pp. 596–616.

44. Max Tegmark, "The Importance of Quantum Decoherence in Brain Processes," *Physical Review E* 61 (1999): 4194–4206.

45. For a discussion with references, see *The Unconscious Quantum: Metaphysics in Modern Physics and Cosmology* (Amherst, NY: Prometheus Books, 1995), ch. 10.

46. Taner Edis, "How Gödel's Theorem Supports the Possibility of Machine Intelligence," *Minds and Machines* 8 (1998): 251–62; *Science and Nonbelief* (Amherst, NY: Prometheus Books, 2008), pp. 98–103.

47. Stenger, *The Unconscious Quantum*, p. 286.
48. Stenger, *Has Science Found God?* pp. 120–21.

9. THE WAY OF NATURE

The Tao is empty
When utilized it is not filled up
So deep! It seems to be the source of all things

—Verse 4 of the Tao Te Ching

THE AXIAL SAGES

In *The Great Transformation*, religious historian Karen Armstrong tells the story of the great teachers of ancient times who dramatically transformed human thinking. German philosopher Karl Jaspers had termed their period, starting around 900 BCE and ending about 200 BCE, the *Axial Age*, because of it being a pivot around which humans turned to see a new perspective of themselves and their relation to the world:

> If there is an axis in history, we must find it empirically in profane history, as a set of circumstances significant for all men, including Christians. It must carry conviction for Westerners, Asiatics, and all men, without the support of any particular content of faith, and thus provide

> all men with a common historical frame of reference. The spiritual process which took place between 800 and 200 B.C.E. seems to constitute such an axis. It was then that the man with whom we live today came into being. Let us designate this period as the "axial age." Extraordinary events are crowded into this period. In China lived Confucius and Lao Tsu, all the trends in Chinese philosophy arose.... In India it was the age of the Upanishads and of Buddha; as in China, all philosophical trends, including skepticism and materialism, sophistry and nihilism, were developed. In Iran Zarathustra put forward his challenging conception of the cosmic process as a struggle between good and evil; in Palestine prophets arose: Elijah, Isaiah, Jeremiah, Deutero-Isaiah; Greece produced Homer, the philosophers Parmenides, Heraclitus, Plato, the tragic poets, Thucydides and Archimedes [and the scientists Thales and Democritus]. All the vast development of which these names are a mere intimation took place in those few centuries, independently and almost simultaneously in China, India and the West.[1]

It must be mentioned that not all historians accept the notion of an Axial Age because it implies some kind of force acting in history.[2] However, historian Steve Farmer suggests this coincidence may have been the result of the expanded availability of lightweight reading materials that occurred at about that time.[3]

So with some caution, let us look at how Armstrong has interpreted this period. Note that she frequently uses the terms "spiritual" and "mystical," which have supernatural implications. I take the attitude that nothing supernatural need be involved.

By 1000 BCE most societies had developed beliefs in a myriad of gods who dwelled in both the heavens and Earth, and who were responsible for everything that happened. People engaged in elaborate rituals designed to propitiate the gods, with animal sacrifice a universal practice.

The axial sages taught a radical philosophy in which people were to seek reality inside themselves rather than by looking outward to the heavens. The new view developed within Hinduism and Buddhism in India; Confucianism and Taoism in China; monotheism in Israel; and philosophical rationalism in Greece. Armstrong identifies the primary sages as Confucius, Lao Tzu, Buddha, Jeremiah, Mencius, Socrates, Euripides, and unnamed mystics of the Hindu Upanishads. Armstrong enthuses: "The Axial Age was one of the most seminal periods of intellectual, psychological, philosophical, and religious change in recorded history; there would be nothing comparable until the Great Western Transformation, which created our own scientific and technological modernity."[4]

We caught a glimpse of axial thinking in our study of suffering and morality in chapter 6. According to Armstrong,

> All the sages preached a spirituality of empathy and compassion; they insisted that people must abandon their egotism and greed, their violence and unkindness. Not only was it wrong to kill another human being, you must not even speak a hostile word or make an irritable gesture. Further, nearly all the axial sages realized that you could not confine your benevolence to your own people: your concern must somehow extend to the entire world.[5]

In line with the caveat I mentioned above, I see no basis for using the words "spirituality" and even "religion" in describing the nature of the movement. My personal study of the main teachings of the axial sages indicates little emphasis on metaphysics or theology, at least in the East. While they still valued ritual, without animal sacrifice, to help put one in the proper state of mind, morality was at the heart of their teaching. What they called "God," "Nirvana," "Brahman," or the "Way" was reached by living a compassionate life. You did not begin your quest by being "born again" and dedicating yourself to some absolute, preestablished notion of deity.

THE INNER FIRE OF ATMAN

Let us now look at how the Axial Age developed specifically in various regions according to Armstrong. Around 900 BCE Hindu rituals dramatically changed. Previously those elaborate ceremonies had focused on sacrificing to external gods and warding off evil spirits. The goal was to get the gods' help in achieving material goods, cattle, wealth, and status. There was little or no self-conscious introspection.[6]

Then suddenly, at least by historical standards, ritual reformers began redirecting the rites toward the creation of *atman*, the self. Atman was not the same as what is called the soul in the West. It was not wholly spiritual and was identified as the inner fire of the human being. Soon it was realized that a person did not have to participate in any external liturgy at all but could achieve the same result with solitary meditation. Armstrong concludes: "The Axial Age of India had begun. In our modern world, ritual is often thought to encourage a slavish conformity, but the Brahmin ritualists had used their science to liberate themselves from the external rites and the gods, and had created a wholly novel sense of the independent autonomous self."[7] The earliest Hindu scriptures, the Rig-Veda, were written at about the time the Aryan tribes arrived in India around 1500 BCE. They were inspired by endless migration and warfare. With the turning inward of the Axial Age, scriptures called the Upanishads were written to accommodate the new view. They eventually achieved the same status as the Rig-Veda. The focus of the Upanishads was the atman and identified with Brahman, the ultimate, unchanging reality of the Rig-Veda. Armstrong explains, "If the sage could discover the inner heart of his own being, he would automatically enter the ultimate reality and liberate himself from the terror of mortality."[8]

The earliest Indian sage Armstrong mentions by name is Yajnavalkya, the personal philosopher of King Janaka of Videha from around the late seventh century BCE. Armstrong credits Yajnavalkya with a proposal that would become a central belief in every major religious tradition—that an immortal spark exists at the core of the human person that gives life to the entire cosmos.

AXIAL JUDAISM

The Axial Age remarkably reached its peak in India at about the same time, the sixth century BCE, as it did in China, Greece, and the Middle East.

In 586 BCE Nebuchadnezzar II captured Jerusalem and carried many, though not all, Jews off into exile in Babylon. He left behind a desolate ruin that is described in the book of Lamentations. Having lost everything, the people of Israel created a new vision out of the experience of grief, loss, and humility. This was their axial moment.[9]

Leading them to a new perspective was the prophet Jeremiah, whose very name is used to represent a person of exaggerated pessimism. But his pessimism was not exaggerated. It was accurate. He demanded that people see things as they really are, that they face the truth. "Instead of denying his suffering, Jeremiah presented himself to the people as a man of sorrows, opening his heart to the terror, rage, and misery of his time, and allowing it to invade every recess of his being. Denial was not an option. It could only impede enlightenment."[10]

Having lost everything that was external, the people of Israel began to turn inward. The exiles in Babylon also reoriented their view. The book of Job may have been written there. Since YHWH behaves so badly in that story, this may indicate that the exiles were losing faith in him. Other parts of the Old Testament may also have been written then—almost certainly Genesis, which is at least partly based on the Babylonian creation myth.

AXIAL GREECE

In 585 BCE (a year before the destruction of Jerusalem, but unconnected in any way) Thales of Miletus, a Greek colony on the coast of Asia Minor, supposedly predicted an eclipse of the sun that, according to Herodotus, put an end to a war between the Lydians and the Medes. Armstrong gives 593 BCE as the date, but I don't believe this is correct.[11] Working backward, astronomers calculate that a total eclipse of the sun did occur in Asia Minor on May 28, 585 BCE, on our current calendar. Historians, however, debate whether the prediction was ever actually made. Assuming it was, the eclipse prediction was based on the assumption that such rare heavenly events occur on a natural, orderly basis and were not the product of some god's whim—a major advance in human thinking.

Whether or not he predicted the eclipse, Thales made the first recorded attempt to explain the universe without recourse to the supernatural. This makes him the first scientist in recorded history. While the Egyptians and others had advanced technologies much earlier, they thought more like engineers than scientists—about building devices and edifices—while Thales thought about how the universe was constructed. He proposed that everything was "born from water." While he was wrong about the basic ingredient, he was still thinking like a modern particle physicist.

In any case, immediately after Thales Greek science began to flourish—and might have continued to do so to the present day except for an interlude called the Dark Ages, brought about by the dark forces of Christianity. Around 480 BCE Anaximenes proposed everything was made of air and Heraclitus said that everything was fire. Around 435 BCE Empedocles proposed the idea of basic *elements*, which he identified as *earth*, *air*, *fire*, and *water*. About 270 BCE Democritus and Leucippus had the remarkable insight that has stood to this day, that everything is composed of tiny elementary bodies they called *atoms*.

While Greek science was no doubt another turn on the axis, like the turn made by the Jews at the time of the exile in Babylon, it was also far from the inward-looking revolution that was taking place in India and China. If anything, it introduced the outward-looking attitude that characterizes science to this day.

SAMKHYA

During that same remarkable sixth century, a new philosophy emerged in India called *Samkhya* that deviated from the Upanishads. The founder is believed to be a sage named Kapila, about whom little is known. As Armstrong describes it,

> Samkhya was an atheistic philosophy. There was no Brahman, no *apeiron* [(Greek): the indefinite original substance of the cosmos in the philosophy of Anaximander], and no world soul into which everything would merge. The supreme reality of the Samkhya system was *purusha* (the "person" or "self").[12]

But this self was not the Purusha figure in the Rig-Veda, the atman of the Upanishads, or the Western soul. It is difficult to grasp what exactly it was supposed to be. Obviously its existence was not the result of objective scientific observation. Rather, it was a kind of inner light glimpsed during meditation that indicated another, more absolute self dissociated from the "natural" realm of both matter *and mind*.

Here again we see a way being sought to escape suffering. The ego was the problem. It traps us into a false sense of self, whose eternal survival is a delusion. Once we realize the ego is not our real self, we can achieve *moksha* ("liberation").

This new self-awareness was not self-indulgence, which was the ego holding us in thrall. Samkhya liberates us from our ego-based existence.

The new state of being is not divine, not supernatural. Anyone can achieve it. The existence of suffering was not denied. It is part of nature.

Liberation, however, was extremely difficult to achieve. Meditation was not always enough. As a result, the discipline of yoga was developed. This was not the yoga practiced so widely today. It was not an aerobic exercise designed to help people relax and feel better about themselves. The original yoga was a systematic assault on the ego in which the aspirant had to learn to abolish the normal consciousness with all its delusions and replace it with *purusha.*

BUDDHA

As we saw in chapter 6, the principle philosophy of Buddhism is to eliminate suffering by achieving the state of nirvana in which the chain of lives terminates and the self becomes nonexistent. This was based on Samkhya, but Sidhärtha Gotama, called the Buddha, who lived toward the end of the fifth century, was skeptical of its metaphysical doctrines. He saw purusha as manufactured during a yogic trance. He had no problem with yoga; he just did not think that it opened the door to a new reality.

In fact, Gotama developed his own version of yoga. While tradition has him sitting under a bodhi tree and gaining the insight in a single meditation that liberated him from the cycle of birth and rebirth, it probably took years for him to develop the Noble Eightfold Path. But, according to tradition, he eventually achieved the peace of complete selflessness, an inner haven in which a person who put his regimen into practice could experience a profound serenity in the midst of suffering.

While nirvana was a transcendent state, it was not associated with the Hindu Brahman. Nirvana was the extinction of greed, hatred, and delusion, an ultimate refuge.

But achieving his own personal nirvana, the Buddha realized he could not ignore the rest of the world. That would violate an essential dynamic of his *dhamma*, his way of salvation. To live morally was to live for others. For the next forty-five years the Buddha traveled up and down the Ganges plain, bringing his teachings to the people. He assembled a huge following of disciples who practiced his methods and who, in turn, brought them to others.

The key doctrine was *anatta*, the elimination of self. The Buddha rejected all questions about cosmology or gods. Knowing the answers would not end suffering and pain. They were of no help in negotiating the *dhamma*.

To those who could not travel the road, who found they were too attached to self, the Buddha did not scold them but asked them to consider that others might feel the same, and so a person who loves the self should not harm the self of others—a version of the Golden Rule. Since only monks had the time necessary to spend meditating to achieve full selflessness, laypeople could still understand their own selfishness and empathize with others. They could go beyond the excesses of ego and learn the true value of compassion.[13]

So the Buddha showed that it was possible to live in this world of pain at peace, in control, and in harmony with one's fellow creatures.

CONFUCIUS

The Axial Age in China began with the man we know as Confucius (d. 479 BCE). Unlike the other axial sages of the East, Confucius was not a solitary ascetic but a man of the world. He did not practice introspection or meditation. Rather, he enjoyed a good dinner, fine wine, a joke, and stimulating conversation.[14]

We know him from the Analects, hundreds of short, unconnected remarks with no clearly defined vision put together by disciples long after his death. You must search for meaning between the lines. That search becomes continuous and endless, never reaching a final conclusion.

But Confucius (c. 500 BCE) was nonetheless an axial philosopher since he profoundly disagreed with the conventions of the times. China was in great disorder and Confucius called for a return to an earlier time when people supposedly lived together harmoniously and pursued the "Way of Heaven." The true gentleman was not to be a warrior but a scholar, following strict rules of correct behavior as prescribed by his place in society. Women, of course, received little attention.

However, Confucius also broke with ancient tradition, which focused on heaven and performing sacrifices to gain favor of gods and spirits. Instead Confucius concentrated on this world, something we knew about.

Like the other axial sages he dismissed metaphysics and theological discourse. When pressed to tell how to serve the gods, he replied, "Till you have learned to serve men, how can you serve spirits?" When asked what the life of the ancestors was like, he responded, "Till you know about the living, how are you to know about the dead?"

So Confucius did not teach about god and spirits but how to be good here on Earth. His Way was not the Way to heaven, but the Way for its own sake. It led to a state of transcendent goodness. The Way was not meditation, however, but ritual. Once again the ego is singled out as the source of human pettiness and cruelty. The rites were designed to suppress the ego. One still sees this today in the Chinese family with the son serving the father, all with proper demeanor. The father cannot behave as some superior lord of the household but must return absolute respect to the son.

Confucius taught the Golden Rule pretty much as we recognize it today. Indeed, the Way was nothing but a dedicated, ceaseless effort of doing your best for others.[15]

LAO TZU

The Tao Te Ching (Classic of the Way of Virtue) is one of the most translated books of all time. It consists of eighty-one short, enigmatic poems, most of them less than a page long.[16] The Chinese sage Lao Tzu (c. 500 BCE?), believed to be a slightly older contemporary of Confucius, is the traditional author of the Tao. However, the poems may have been composed by others writing with this pseudonym. According to some modern scholars, Lao Tzu is entirely legendary.[17]

Taoism did not begin with the Tao Te Ching but had been present in Chinese culture for thousands of years. Also, it was not written for individuals but as advice to rulers, although it has been adopted by self-help gurus in the West as a guide for living life with grace, peace, and joy. Besides reading the verses myself and studying various scholarly commentaries, I have listened to the audiobook of *Change Your Thoughts, Change Your Life: Living the Wisdom of the Tao* by the well-known author and frequent *Oprah* guest, the self-help guru Wayne Dyer.[18]

In the Tao Te Ching we again hear the now familiar refrain of the axial sages—that one must suppress one's ego and live in harmony with nature. The "Tao" itself is the unseen reality behind everything that is the same thing as the "Void"—nothingness. I might claim a connection here with the cosmology of nothingness I presented in chapter 7, but I could hardly prove it.

The Tao Te Ching poems often seem to make no sense and may have been designed that way to encourage meditation, a technique picked up later by Zen Buddhism. For example, the Tao is also nameless (v. 32). It is constant in nonaction, yet there is nothing it does not do (v. 37). It is empty. It is not God. It is the predecessor of God (v. 4).

Dyer, as do Armstrong and others, refers to all of this as "spiritual" and "metaphysical." But when it comes down to the actual recommendations for negotiating the Way, nothing supernatural comes into play. For example, as in all forms of Eastern meditation and in contrast to Western meditation, one does not pray to some deity. The Chinese already had a form of yoga and Lao Tzu did not lay down any particular practices. However, he must have had meditation in mind when he tells his listener to experience the "emptiness" that is the Tao. The listener thus returns to the authentic humanity before civilization, which had introduced a false artifice into human life. By interfering with nature, human beings had lost their Way.[19]

The message that I kept hearing repeated over and over again in Dyer's commentary was that the individual should be forever passive. Go with the flow of nature. Take no actions but let the actions take care of themselves. Ignorance is knowledge. Weakness is strength. Submission is hardness. By putting your person last, you come in first. Easy to say for a millionaire living on the beach on Maui, where he can smell the plumeria and contemplate the sunset.

The teachings of Lao Tzu were and still are impractical as a guide to life in the external world. He had hoped to end the continuous violence of what Armstrong called the Warring States by the people joining together peacefully under a compassionate ruler. Instead China was unified by the achievement of empire—the military conquests of the state of Qin, which became China as we know it. However, the inner life of the Tao matched that of the other sages, and one way to meditate is to read a verse of the Tao Te Ching and then sit quietly thinking about it and nothing else.

THE RETURN OF THE SELF

Armstrong summarizes:

> The Axial sages put the abandonment of selfishness and the spirituality of compassion at the top of their agenda. For them religion *was* the Golden Rule. They concentrated on what people were to transcend *from*—their greed, egotism, hatred, and violence. What they were going to transcend *to* was not an easily defined place or person, but a state of beatitude that was inconceivable to the unenlightened person, who was still trapped in the toils of the ego principle.[20]

Of course, there have been many other religious figures who have taught similar principles of selflessness and looking inward. We can find them in the Gnostic gospels discovered in Nag Hammadi, in the Gospel of John, and in the life of poverty lived by St. Francis of Assisi (d. 1226). The Spanish mystic St. John of the Cross (d. 1591) taught *kenosis*, the emptying of the mind of all content and imagery. Members of the Judaic Kabbala sect meditate to cleanse their minds of negative influences. Sufi Muslims also meditate in an attempt to transcend from the personal to the divine.

These were all mystics who never doubted that they were engaged in a supernatural activity. This is far less evident in the case of the Eastern sages. What's more important, the state of mind that all the various techniques achieved was identical to and perfectly compatible with a purely materialistic mind.

An objective look at the religions of the world today, including those founded by the axial sages, shows that almost all are marked by the return of the self as the center of religious consciousness. These religions all have priests, nuns, and monks who more or less follow the teachings of their revered sages and saints. But the great bulk of adherents do not practice those teachings.

Hindus pray to the gods for selfish needs and look forward to future lives as Brahmins. Those who are already Brahmins show no regard for those of lesser castes. Indeed, members of each caste look down on the ones below. I have known many Hindus from India. Most have treated me with kindness and respect. Yet I have watched them with other Hindus and, without exception, I have seen no hint of compassion for people of lower caste.

Similarly, most Buddhists, Taoists, and Confucians look out for their own personal welfare, both in this world and the next. In the West, this emphasis on the ego is even more pronounced. Jews continue to regard themselves as the chosen people of God. And who could be more self-centered than Muslims and Christians, who believe that for a few simple duties for a short period on Earth they will live forever in perfect bliss?

QUANTUM SPIRITUALITY

Although the followers of the axial sages have since attached supernatural elements to their teachings, the sages themselves largely disregarded any talk of cosmic power. This is in striking contrast to what is taught today by gurus of "quantum spirituality," that we can control our own reality and that our consciousness tunes into a field of energy that pervades the universe.

I discussed this form of new spirituality in my previous book, *Quantum Gods: Creation, Chaos, and the Search for Cosmic Consciousness*.[21] Let me just summarize my observations.

In 1975 physicist Fritjof Capra authored a best seller called *The Tao of Physics* in which he proposed a connection between Eastern philosophy and quantum physics.[22] Capra's main theme was that quantum physics shows that the universe is one connected "whole." By searching through the voluminous literature on Eastern philosophy, he was able to mine quotations that vaguely supported his thesis. His use of "*Tao*" in the title of the book refers to its meaning as "Way" rather than any specific reference to the Tao Te Ching, which he rarely quotes. In fact, very few specifically axial teachings appear in Capra's selection of concepts from Eastern philosophy that are alleged to connect to physics. Capra does not tell us to turn inward. Rather, he tells us to look outward to the vast universe and imagine that we are a grand part of its reality.

A few years before Capra, an Indian yogi who called himself Maharishi Mahesh Yogi gained priceless publicity from a brief involvement as guru to the Beatles and other entertainment stars who visited with him in India. Maharishi proceeded to build an empire in America and Europe, teaching his mode of meditation called *transcendental meditation* (TM). This form of meditation was easy to learn and became very popular, no doubt because meditation of any kind at least briefly improves a person's sense of well-being.

Trained in physics, the Maharishi cleverly raised the credibility of TM, at least among nonphysicists, by claiming scientific support. He did this in two ways. First, he hired researchers to study the effect of meditation on various body parameters. Many overblown claims were made for the health benefits of TM, which in fact were little more than the aforementioned feeling of well-being, some temporary lowering of blood pressure, and the ubiquitous placebo effect that accounts for most of the purported benefits of alternative medicine.

Second, and far more dubiously, the Maharishi asserted that TM put one in contact with the "grand unified field" of consciousness that existed throughout the universe. Physicists had just developed the standard model of particles and fields, which I briefly discussed in chapter 7. One of its key ingredients was the unification of electromagnetism with the weak nuclear force. The next step being eagerly pursued by physicists was what was called the *grand unified theories* (GUTs), which united the new "electroweak" force with the strong nuclear force. The Maharishi seized on this as the cosmic force that comes into play during transcendental meditation.

As it turned out, the simplest GUT made a very precise prediction that the proton is unstable. This instability is very tiny or else none of us would be here. It was calculated from the minimal theory that one proton should decay on average every 10^{31} years. Now, we obviously can't wait that long for a single proton, but if we watch at least 10^{31} protons for a year or more, we should see some decay.

This was within the capability of existing technology, and several major experiments were quickly mounted. After a few years of seeing no decays, they were able to falsify the minimal GUT. More complicated theories were considered, and are still being tested, but for the most part, GUT has largely been abandoned in favor of string theory and other schemes that seek unification of all forces including gravity, which is not a part of any GUT.

One of the Maharishi's early disciples was physician Deepak Chopra, who saw the opportunity for greater personal success by going out on his own. Chopra and others in what came to be called the "New Age" of spirituality exploited some of the wild claims that were being made by popular writers, including some physicists, about the implications of quantum mechanics. I examined these claims in my 1995 book, *The Unconscious Quantum*, and have updated these in *Quantum Gods*.[23]

The New Age authors interpreted quantum mechanics as implying that our conscious minds are able to control reality. Chopra specifically claimed that, with the right thoughts, we could heal ourselves and delay aging.[24] He continues to age with the rest of us.

More recently, the notion that "we make our own reality" was rediscovered and exploited in two highly successful documentary films and their spin-offs in terms of books, CDs, and other paraphernalia: *What the Bleep Do We Know!?* directed by William Arntz,[25] and *The Secret*, authored by Rhonda Byrne. Nothing in these films and books was not already being claimed decades earlier in the period I called the New Age. Since the New Age is no longer new but its ideas are not about to go away, I will refer to the whole phenomenon as *quantum spirituality*.

The Secret is particularly dramatic in its claims, besides the fact that its principles were kept a secret for centuries by the usual cast of characters, such as Leonardo da Vinci and Albert Einstein, until discovered by author Rhonda Byrne's daughter. Readers are informed, over and over again, that all they need to do in order to have everything they want in life—health, beauty, power, wealth, love—is to simply *think* they want it. It is guaranteed! You must not allow any negative thought ever to enter your head. Meditation methods have since become available to help you do this.

If you do not achieve your desires, it is your own fault for not thinking positively. The starving people of Darfur just have to stop thinking so negatively. Look at what happened to the Jews in the Holocaust!

While all these promoters of quantum spirituality claim a kinship to Eastern philosophy, we see that, just as in the case of the god-centered religions, the emphasis is hardly on the suppression of ego. What could be more self-centered than wanting to control reality with your own thoughts?

MAKING YOUR OWN REALITY

In my two books on this subject I considered in some detail the claims that we make our own reality, along with Fritjof Capra's assertion that quantum mechanics implies the universe is one interconnected whole. Both trace back to a very simple notion that has been grossly misrepresented, by both popular writers and some credentialed physicists as well: the *wave-particle duality*.

Early in the twentieth century, phenomena that we had previously described as wavelike, such as light, exhibited particle-like behavior. The particles of light were named *photons.* At the same time, phenomena that we had previously described as particle-like, such as electrons, were observed to also behave like waves. Furthermore, it seemed that whether an object was a wave or a particle depended on what you decided to measure. If you decided to measure a particle property such as position, then the object was a particle. If you decided to measure a wave property such as wavelength, then the object was, presumably, a wave. It was all under the control of human consciousness.

The object might be a pulse of light from a galaxy billions of light-years away. But whether it was a particle (photon) or a wave depended on the observer's *conscious* decision, made billions of years after the object left its source, on what property to measure. Thus the conclusion, promoted by Deepak Chopra and other quantum spirituality gurus, is that we decide for ourselves the nature of reality by an act of conscious will.

An implication, which the gurus do not usually emphasize because it is so ridiculous no one will believe it except their most spaced-out disciples, is that our mind actually reaches back 13.7 billions of years in time and reaches out 46 billion light-years in space (the current size of the universe) to specify the nature of every body in the universe. To avoid an anticipated question from science-savvy readers, let me explain where the size of 46 billion light-years comes from. While the speed of light limits our observations to 13.7 billion light-years, because the age of the universe is 13.7 billion years, the universe that began that many years ago has expanded beyond our horizon to 46 billion light-years in size. The speed of light does not limit the range of conscious thoughts, however, which can act instantaneously throughout space—at least, if you believe quantum spirituality. The trouble is, it's almost as hard to believe as a virgin giving birth to someone who thirty years later rose from the dead.

Anyway, many people seek this cosmic consciousness to replace the archaic god-centered religions that remain from humanity's childhood. As Jesus said, "Except ye be converted, and become as little children, ye shall not enter the kingdom of heaven" (Matt. 18:3, KJV). A child cannot be expected to understand quantum mechanics, but, then, neither do Chopra and the quantum spiritualists.

As explained above, quantum spirituality, including Capra's holistic universe, depends on the wave-particle duality. If you measure a particle property of an object, then you have consciously made that object a particle. If you measure a wave property of an object, then you have consciously made that object a wave.

EVERYTHING WE OBSERVE IS PARTICLES

Let us do an experiment to measure the wavelength of light. We shine light on an opaque surface that contains two narrow parallel slits. Thomas Young first performed this experiment in 1800 and discovered the wave nature of light by observing a wave interference pattern on a screen behind the slits.

Today we can easily modernize the double-slit experiment by placing on the screen an array of small detectors of sufficient precision that they are capable of registering individual photons, the particles of light. We find the result illustrated in figure 9.1. We see that each individual photon triggers a single detector, whose position gives the photon's position, verifying it as a particle—even though we set out to measure the wavelength! Yet, the interference pattern emerges statistically as we watch the number of photons increase.

What we observe is exactly what physicists have been saying about quantum mechanics since the 1920s. Those who have quantum snake oil to sell have studiously chosen not to listen. The wave that is associated with particles is not a property of individual particles but the statistical property of an ensemble of many particles. Mathematically it is represented by a complex number called the *wave function*, whose amplitude squared gives the probability of finding a particle in a unit volume at a particular position in space.

In short, there is no wave-particle duality—just as there is no mind-body duality, which is a connection between dualities that is often made. The waves that are associated with microscopic and submicroscopic phenomena are mathematical entities that appear in theories and are never measured directly. Nothing is actually waving. In fact, the wave function in general is a wave in name only. Not every wave function has the functional form of a wave.

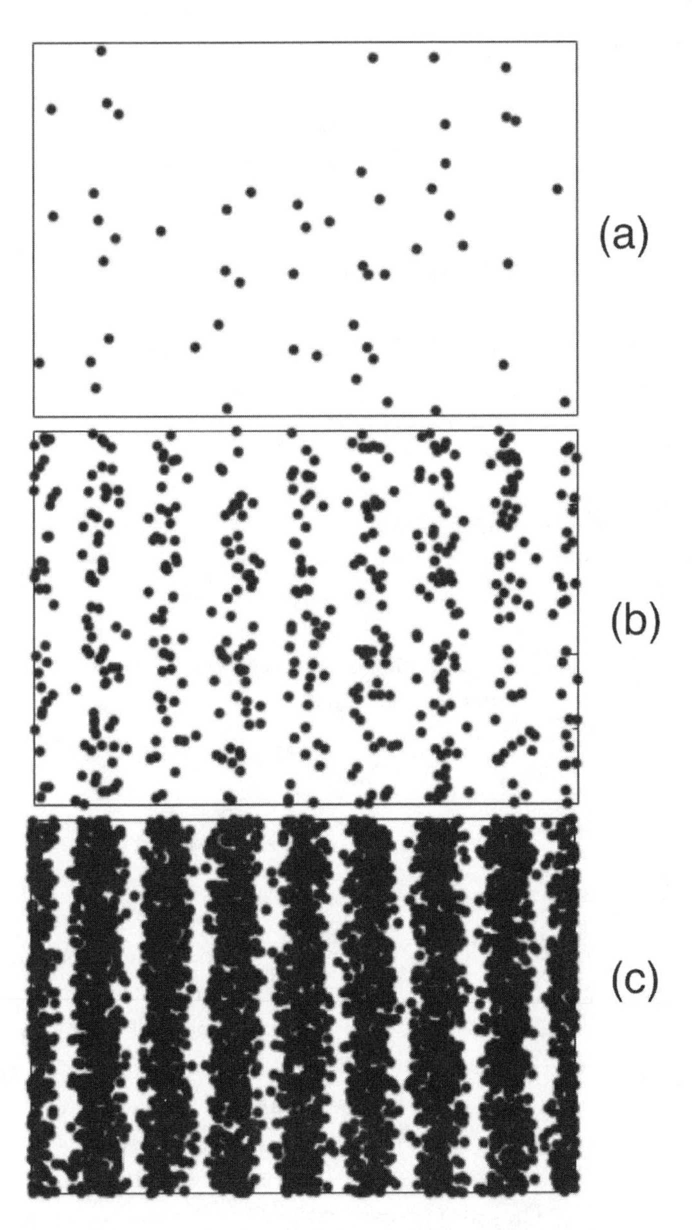

Fig. 9.1. The photon hit pattern in double-slit experiment. With just a few hits (a), we see what looks like a random pattern of localized particle hits. But as the number of hits increases in (b) and (c), the familiar double-slit wave interference pattern emerges as a statistical effect.

SAM HARRIS AND SPIRITUALITY

Sam Harris, whose *End of Faith* best seller pioneered New Atheism, was severely criticized by other atheists and freethinkers for introducing Eastern "spiritualism" or "mysticism" into an otherwise atheistic view of the world and humanity. His own severe criticisms of religion were aimed at the Western varieties, particularly Christianity and Islam. He urged a return to the "wisdom of the East."

At first I reacted negatively as well, especially to his frequent use of the terms "spiritual" and "mystical," which to most people suggest something supernatural—a reality beyond matter. He writes,

> For millennia, contemplatives have known that ordinary people can divest themselves of the feeling that they call "I" and thereby relinquish the sense that they are separate from the rest of the universe. This phenomenon, which has been reported by practitioners in many spiritual traditions, is supported by a wealth of evidence—neuroscientific, philosophical, and introspective. Such experiences are "spiritual" or "mystical," for want of better words, in that they are relatively rare (necessarily so), significant (in that they uncover significant facts about the world), and personally transformative.[26]

Although he does not use the term, we see he is reiterating the teachings of the axial sages.

I have no quibble with his giving this phenomenon scientific credibility by referring to "evidence." Certainly enough people have undergone these kinds of experiences to take the existence of the experiences themselves as scientific facts. Certainly many people have been transformed by these experiences, especially when they take on the character of epiphanies. Less certain, in my mind, is how significant are the "facts about the world" uncovered during these "spiritual experiences" and what exactly are those facts. Sure, they tell us something about how the brain operates, but Harris insists, "They also reveal a far deeper connection between ourselves and the rest of the universe than is suggested by the confines of our subjectivity."[27] This begins to sound like New Age mumbo jumbo to me. However, Harris has clarified his position in an e-mail exchange with physicist Alan Sokal:

> I am referring to facts about our minds—especially to the fact that the ego is a construct and free will (in the deep sense) an illusion. Needless to say there are objective (i.e., neurological) reasons to say the ego is a construct and free will is an illusion, but there are subjective reasons as well. Introspection/meditation can uncover the subjective reasons (not the objective ones).[28]

Sokal asks Harris to elaborate on the statement that mystical experiences "reveal a far deeper connection between ourselves and the rest of the universe." Harris responds:

> This sentence, admittedly, is easy to misunderstand. But I did not mean anything spooky by it. I am simply claiming that the experience of oneself as highly permeable to the world, and ultimately inseparable from it, is more accurate than experiencing oneself as a skin-encapsulated ego. But I am not saying that spiritual experiences of unity with nature (or anything else) allow us to make claims about physics, biology, etc.[29]

Sokal asks if Harris means that "a purely physicalist approach to studying consciousness is likely to miss something, and that first-person investigation is also necessary." Harris replies, "That is exactly what I mean." And he makes clear that he does not mean that through the domain of our subjectivity and the first-person investigation of consciousness, one learns something about the nature of the universe.[30]

Harris has not been alone among materialist scientists and philosophers of mind who are sympathetic to Buddhism and active practitioners of meditation. Susan Blackmore is an atheist psychologist very active in skeptic circles whose definitive studies of psychic phenomena and mystical experiences I have referred to extensively in earlier books. Blackmore has practiced Zen meditation for over twenty years. She makes it very clear, however, that she is not a Buddhist because "I dislike dogma in any form."[31] Nevertheless, Blackmore has found that her Zen practice and scientific work "converge in many ways."

Philosopher Owen Flanagan notes that the "hard problem" in mind science is generally regarded as explaining how mind is possible in a material world. As we saw in chapter 8, while we do not know the answer yet, there does not seem to be any obstacle to a purely material mind. Flanagan argues that, while this problem is hard, the "really hard problem" is finding meaningful and enchanted lives in a material universe.[32] After all, looking at life from a Darwinian perspective we see no sign of purposeful action. We tell ourselves stories about living a meaningful life, but Flanagan wonders what insight that gives us.

Flanagan suggests that Buddhism offers a form of "personal salvation" that is especially attractive to secular naturalists, provided the doctrine of karmic rebirth is excluded.[33] The bodhisattva in Buddhism is defined as a person aiming at embodying or realizing the four divine abodes: compassion, loving-kindness, sympathetic joy, and equanimity. Flanagan says, "Anyone can, and everyone should, take the bodhisattva's vows as a reliable way to live a fulfilled and ethical life."[34]

THE WAY OF NATURE

From what we already know about the brain, matter can plausibly account for all the qualities of human thinking referred to as "spiritual" or "mystical." A recent paper reports on brain scans that support a model in which spiritual experiences are related to *decreased* activity of the right parietal lobe in the brain.[35] In this model, by minimizing right parietal functions using meditation, individuals experience the lessened sense of self and heightened feeling of being one with the universe that is called "transcendence." The same effect occurs when the same region of the brain is injured, strongly indicating that this is a material phenomenon.

It is hard to see how by using *less* of our brains we can learn *more* about the universe, so this experience is most likely all in our heads. Further evidence that meditation does not lead to new knowledge about the world outside our heads is provided by the fact that the methods of science and reason that I am urging everyone to take a stand for in this book did not arise in the East. They obviously were not to be discovered by meditating and looking *inward.* They were also not planted in our brains by evolution (or God). Rather, they have been learned from the experiences of our *outward* observations of the world around us. Maybe, as many Christian apologists claim, Western religions helped science develop by their own looking outward for God. However, I am not ready to give religion too much credit since science goes back centuries before Jesus to the axial age in Greece.

Nevertheless, the resulting state of the brain achieved by meditative practice could still transform a person's life by altering how he thinks about himself and his place in the scheme of things. In particular, if the self is a construction of the brain it can be deconstructed by the brain.

Let us then consider the implications for human life of the atheistic view that everything is matter and nothing more, with the addition of this new insight about looking inward to reduce our unhealthy self-centeredness. Atheism offers no promise of salvation or eternal life. This life is all we have. Most people consider that depressing and unappealing. However, all they have to choose from otherwise are the unlikely prospects of eternal life provided by religions and other spiritualities. Suppose we atheists are right after all? What can be made out of that?

Atheistic materialism means we humans are 100 percent matter, without an ounce of any vague substance called spirit, soul, or living force. It means we humans are 100 percent natural, without a connection to any supernatural agency called God, Allah, or cosmic consciousness. It also means we are free to live our lives as we wish, without any priest or god telling us what to do and think. And since we have no other lives, we live this one to the fullest.

This does not lead to the breakdown of society so feared by theists. The observable facts are that atheists are at least as moral as theists, if not more so because they rely on their own consciences and not what someone else tells them is right or wrong. The atheist must actually take responsibility for his own morality. He can't say, "God told me to do it." And, as we will see in the next chapter, the least religious nations are the happiest and healthiest by every measure.

If you accept atheist materialism, then you have to learn to live with the conclusion that human consciousness and self-awareness reside in a purely material brain and nervous system. They may even be a trick the brain plays on us anyway, without having much to do with reality at all. Most of our brain's decision making is done subconsciously *before* we are aware of thinking that some entity known as "I" made the decision. In any case, all thoughts will cease when the brain and nervous system stop operating and begin the process of rejoining the dust of Earth from which they arose. This is a terrifying prospect for many and I am not offering it as an attractive substitute to eternal life. I don't expect to convert a single believer to atheism by this argument. Here I am talking to those who have already recognized the undoubted fact that there is no eternal life and I am suggesting a possible way to cope with it.

As we have seen, the sages of the East preached that the way to achieve peace of mind is to turn away from the ego-centeredness that we all seem to develop in childhood. Buddha called it the *Noble Eightfold Path.* Lao Tzu called it the *Way to Virtue.* Confucius called it the *Way of Heaven,* though it was to be practiced on Earth. I will simply call it the *Way of Nature.*

By living the Way of Nature we celebrate the natural and refute the supernatural. We accept the fact that our individual selves will pass along with our generation. But succeeding generations will carry our genes and our ideas forward. Having just passed my seventy-fourth birthday, and not being in the best of health (though I am working to improve that), I have to start thinking about my mortality. All my life I have had my ego boosted by those around me. Growing up in an uneducated immigrant family on the edge of poverty in a working-class neighborhood, I was the first in remembered generations of the family over six feet tall, the first to graduate college, the first to get a PhD, the first to write a book. I always have had people praising me, especially now with my books reaching a wide audience. I have not received 100 percent praise, of course, but enough to keep me feeling good about myself.

I have a great personal life, too, having been happily married to the same woman since 1962. Our two children are settled in happy marriages of their own and are both highly educated and successful. Our four grandchildren are beautiful and intelligent (pictures available on request). What else can a man want? I wouldn't mind continuing it forever. But I can't.

Not having yet practiced any serious meditation, I cannot recommend any one method. I understand that Zen is extremely difficult. TM is easy, but stay away from the hucksters who sell it or any other practice that promises too much. Even yoga teachers, good as they are in getting your body and mind in shape, attach it to spiritual and mystical mumbo jumbo. "It's not so much the body as the spirit," they will typically tell you. What this world needs is an honest, effective, and fully materialistic method of meditation.

So, it is going to be very difficult for me to practice what I preach, which is directed to other atheists as they approach the end of their lives: take up the Way of Nature and achieve a state of mind where the self does not matter and nothingness is approached with peace of mind.

But don't do it too soon! Live life first.

Lives of great men all remind us
We can make our lives sublime,
And, departing, leave behind us
Footprints on the sands of time;

Footprints, that perhaps another,
Sailing o'er life's solemn main,
A forlorn and shipwrecked brother,
Seeing, shall take heart again.
—Henry Wadsworth Longfellow[36]

NOTES

1. Karl Jaspers, *The Future of Mankind* (Chicago: University of Chicago Press, 1961), p. 135.
2. New World Encyclopedia, "Axial Age," http://www.newworldencyclopedia.org/entry/Axial_Age (accessed February 21, 2009).
3. Steve Farmer, "Neurobiology, Primitive Gods and Textual Traditions: From Myth to Religions and Philosophies," *Cosmos, Journal of the Traditional Cosmology Society* 22, no. 2 (2006): 55–119.
4. Karen Armstrong, *The Great Transformation: The Beginning of Our Religious Traditions* (New York: Knopf, 2006), p. xvi.
5. Ibid., p. xix.
6. Ibid., pp. 98–100.
7. Ibid., p. 100.
8. Ibid., pp. 148–49.
9. Ibid., p. 198.
10. Ibid., p. 200.
11. Ibid., p. 223
12. Ibid., p. 225.
13. Ibid., p. 341.
14. Ibid., p. 241.
15. Ibid., p. 247.
16. Derek Lin, trans. and ed., *Tao Te Ching: Annotated & Explained* (Woodstock, VT: SkyLight Paths, 2006).
17. Stanford Encyclopedia of Philosophy, "Laozi," http://plato.stanford.edu/entries/laozi/ (accessed February 21, 2009).
18. Wayne W. Dyer, *Change Your Thoughts, Change Your Life: Living the Wisdom of the Tao* (Carlsbad, CA: Hay House, 2007).

19. Armstrong, *The Great Transformation*, p. 409.

20. Ibid., p. 468.

21. Victor J. Stenger, *Quantum Gods: Creation, Chaos and the Search for Cosmic Consciousness* (Amherst, NY: Prometheus Books, 2009).

22. Fritjof Capra, *The Tao of Physics: An Exploration of the Parallels between Modern Physics and Eastern Mysticism* (London: Wildwood House, 1975).

23. Stenger, *The Unconscious Quantum.*

24. Deepak Chopra, *Quantum Healing: Exploring the Frontiers of Mind/Body Medicine* (New York: Bantam Books, 1989); *Ageless Body, Timeless Mind: The Quantum Alternative to Growing Old* (New York: Harmony Books, 1993).

25. William Arntz et al., *What the Bleep Do We Know!?: Discovering the Endless Possibilities for Altering Your Everyday Reality* (Deerfield Beach, FL: Health Communications, 2005); Rhonda Byrne, *The Secret* (New York: Attria Books, 2006).

26. Sam Harris, *The End of Faith: Religion, Terror, and the Future of Reason* (New York: Norton, 2004), p. 40.

27. Ibid.

28. From an e-mail exchange between Sam Harris and physicist Alan Sokal.

29. Ibid.

30. Ibid.

31. Susan Blackmore, *Ten Zen Questions* (Oxford: Oneworld, 2009); "Zen," http://www.susanblackmore.co.uk/Zen/intro.htm (accessed February 21, 2009).

32. Owen J. Flanagan, *The Really Hard Problem: Meaning in a Material World* (Cambridge, MA: MIT Press, 2007).

33. Ibid., p. 99.

34. Ibid., p. 209.

35. Brick Johnstone and Bert A. Glass, "Support for a Neuropsychological Model of Spirituality," *Zygon* 43, no. 4 (2008): 861–74.

36. From Henry Wadsworth Longfellow, "A Psalm of Life—What the Heart of the Young Man Said to the Psalmist."

THE FUTURE OF ATHEISM

> **I did not lose my faith—I gave it up purposely. The motivation that drove me into the ministry—to know and speak the truth—is the same that drove me out. . . . Opening my eyes to the real world, stripped of dogma, faith and loyalty to tradition, I could finally see clearly that there was no evidence for a god, no coherent definition of a god, no agreement among believers as to the nature or moral principles of "God," and no good answers to the positive arguments against the existence of a god, such as the problem of evil. And beyond all that, there is no need for a god. Millions of good people live happy, productive, moral lives without believing in a god.**
>
> —Dan Barker[1]

THE ORIGIN OF RELIGION

As Daniel C. Dennett makes clear in *Breaking the Spell*, we still do not have a scientific consensus on the origin of religion. A common theory is that religion is somehow built into our brains, that we have a "god gene" or simply a predisposition to look for agents behind all phenomena.[2] If so, then it most likely would have been put there by evolution.

Usually, when we talk about the evolution of a part of the body, we assume that it had some survival value—if not for the individual, then for a gene. However, this is not always the case. Sometimes our body has a part, or a particular structure, that just occurred by accident and wasn't sufficiently harmful to be weeded out by natural selection. The god gene might be one of these. Or, more likely, it does not exist.[3]

The early, tribal precursors to religion were divination and magic. We can imagine a shaman in some ancient forest tribe stumbling across some black powder that exploded with a loud bang and bright flash of light when thrown into a fire. He would be sorely tempted to claim special powers for himself by producing such an explosion in front of the tribe during a ritual.

Shamans were expected to reveal information about the agents who ran the world and to help manipulate them. There was no sharp distinction between natural and supernatural. The morals suitable to small groups—fairness, empathy, tit-for-tat, honesty, reciprocity, and suspicion of outsiders—have plausibly been built into us by social evolution. Several of them have been found in the higher social animals like chimpanzees and wolves. But with the development of agriculture humans began to live in much larger groups. Interactions of trade and commerce were between strangers, and tribal morality was too narrow to regulate these interactions. Leaders sought to impose broader rules of behavior, and supernatural agents were invoked to underwrite these rules. Of course leaders also saw securing their own power as an essential good of the society. It is no accident that the origination of religions in the Middle East coinciding with the development of agriculture was hierarchical with priests and kings acting as conduits for divine authority. Ultimately the kings of Christendom asserted divine right with the help of the Church. And in China, the emperor had long been viewed as the conduit to heaven.

In chapter 5 we saw how, despite the official atheism of the communist state, Stalin normalized relations with the Russian Orthodox Church during World War II. You can bet that did not come without quid pro quos. For example, the Church helped the government suppress rival denominations that were not so cooperative.[4] In centuries past the Russian Orthodox Church had close connections with the Czarist secret police, so it is no surprise that Stalin and now the current Russian strongman, former KGB officer Vladimir Putin, embraced the ROC. So you don't even need a nation and a government of believers to use religion to keep people in line.

Although religion and morality have always gone hand in hand, this does not imply that religion is necessary for morality. Most moral precepts can be interpreted as part of the social contract we make with each other in order to live together in peace and harmony. When we don't, as is often the case, the cause can be attributed to a breakdown of that contract. Religion is often a big player in that breakdown.

A GOD GENE?

It seems to me highly unlikely that there is a god gene or any other mechanism built into the brain to specifically produce religious belief. If there were, then there wouldn't be over a billion nonbelievers in the world. Furthermore, whatever special survival value may have been provided by religion has not yet had time to shape significantly the evolution of the human body. People forget that the timescale of biological evolution is vast by human standards. We have essentially the same bodies our ancestors had 100,000 years ago and our brains have not changed much in at least half that time.[5] However, it may be that the brain had certain systems already built in for other reasons that predisposed it to religious thinking once society developed, as suggested by Pascal Boyer in *Religion Explained.*

Now, that does not mean the social systems could not have evolved over shorter times. The mechanism for social evolution is still debated, but it seems reasonable that it happens. In that case religion may have evolved socially, rather than biologically, because of survival or other value to the society that practiced one form or another—or, accidentally. In an article in *Free Inquiry,* historian Alexander Saxton quotes me as saying:

> [Empirical] evidence does not support the widespread assertion that religion is especially beneficial to society as a whole. Of course, it has always proved extremely beneficial to those in power—helping them to retain that power—from prehistoric times to the [2004] presidential election. But it is not clear how society is any better off than it would have been had the idea of gods and spirits never evolved.[6]

Saxton argues that it is difficult to imagine that religion, at least historically, would not have become so universal to our species had it not contributed adaptively to human survival. I don't disagree, except to dispute that it is all that universal. My statement was that things might have been better without religion. The reason religion won out over nonreligion had more to do with the survival of power elites than the benefit of the masses. In fact, Saxton attributes the connection between war and religion as the mechanism by which religion abetted the process. Religion stimulates war and war stimulates technological progress that, in the long run, enhances the global dominance of the human species.

However, Saxton points out that with the Second World War and the nuclear bomb, along with ecological burnout, things changed. Religion ceased to be adaptive to human survival and became dysfunctional. Instead of putting the blame where it belongs, on all of us, religions seek to place the blame elsewhere—on agents of an Evil Empire. Saxton concludes:

> The truth—the hard core, "get real" kind of truth—is that somewhere down under, by some sort of subliminal awareness, every human being really *knows* that believing in belief (as the song famously tells us about falling in love with love), is "nothing but make believe." The atheist's mission is to nourish the seed beneath the snow; to seek not escape but survival.[7]

After submitting this manuscript a prepublication copy of a book by psychologist Hank Davis titled *Caveman Logic: The Persistence of Primitive Thinking in a Modern World*[8] was sent to me by the publisher. Davis makes the case that we humans are all burdened by delusional thinking, such as belief in the supernatural, because our brains are basically the same as they were in the Pleistocene epoch 100,000 years ago. At that time evolution had produced cognitive "modules" in the brain where we tended to make frequent false identifications of simple phenomena as dangerous, such as a predator from which to flee or prepare to defend against. If these modules demanded high certainty they would occasionally miss a real threat, so the threshold was set very low. Obviously, a lot of false positives are far more conducive to survival than a single false negative. In today's world we do not have the same need to react to any hint of danger, yet we have the same brain modules providing us with numerous false positives where we think we see something that is not there. Science, then, is the method by which we examine phenomena and eliminate, as far as is possible the delusions built into our brains 100,000 years ago.

THE GODS ARE NOT WINNING

In their 2004 book, *Sacred and Secular*, Pippa Norris and Ronald Inglehart take a sociological look at the world trends in religion and secularism.[9] As they point out, "The seminal social thinkers of the nineteenth century—Auguste Comte, Herbert Spencer, Émile Durkheim, Max Weber, Karl Marx, and Sigmund Freud—all believed that religions would gradually fade away in importance and cease to be significant with the advent of industrial society."[10] However, religion is still very much with us over a century later and many observers have drawn the opposite conclusion, that secularism is fading away: "After nearly three centuries of utterly failed prophesies and misrepresentations of both present and past, it seems time to carry the secularization doctrine to the graveyard of failed theories, and there to whisper '*requiescat in pace*.'"[11]

Looking at the data, however, secularism is far from dead. In fact, it is the fastest-growing "belief system" in the world. The number of nonreligionists in the world grew from 3.2 million in 1900 to 918 million in 2000. The rate of agnostics and atheists is now expanding at 8.5 million converts each year. They have moved from 0.2 percent to 15 percent of the population in that time.[12]

Even in the United States the religion trend is slowly and surely downward. From 1978 to 2008 church membership dropped from 70 to 65 percent. Bible literalists decreased from 40 percent to 30 percent. Bible skeptics grew from 10 to 20 percent. Mormonism is the exception, but with its 12 million adherents it is still well below nonbelievers.

As few as one in four Americans are actually in church on a typical Sunday, with only a few percent in the megachurches. The Southern Baptist Church, the largest born-again sect, is baptizing at about the same rate as fifty years ago, when the population was half what it is today.

Since 1900 Christians have remained at a steady level of one-third of the world's population, with no growth evident. Neither are Hindus experiencing growth, at a seventh of the total. Paganism, including modern versions such as Scientology and New Ageism, has contracted by half.

In a hundred years Islam has risen from one-eighth to one-fifth of the world and is predicted to be a quarter by 2050. This is the result of the large birthrate among Muslims rather than converts. The result is a rapidly growing proportion of Muslims worldwide, little or no growth in Christians and Hindus, and at the same time a rapidly growing proportion of nonbelievers. The losers are other believers. Nonbelievers now form majorities in Europe, Australia, New Zealand, and Japan.

The expected revival of religion in Russia that was predicted with the end of communism has not materialized. Despite the support given by the government to the Russian Orthodox Church, which was mentioned above and in chapter 5, only 25 percent of Russians believe in God. Faith is even declining in previously very Catholic Poland.

UNDERSTANDING THE DATA

Norris and Inglehart have pinpointed where the sharp decrease in religiosity is taking place. In the period 1917 to 1984, religiosity remained largely unchanged in agrarian societies, where the economies are based on agriculture, and in industrial societies, where the economies are based on manufacturing. However, it has fallen sharply in "postindustrial" societies, where the economies are based on services. These are societies where at least two-thirds of the gross domestic product arises from services.

These authors also confirm the observations of others that a strong correlation exists between religiosity and income, both individually and by nation. When you plot the religiosity of a nation against its per capita gross domestic product you see a strong correlation in which religiosity falls off as a nation is increasingly wealthy (see Fig. 10.1). A striking exception is the wealthiest nation of all, the United States. In America religiosity is on par with the far poorer nations of Mexico and Chile. This is called the *American anomaly*.

But how anomalous is America? When we compare religiosity with income inequality, not only does the United States remain high, but so do Ireland, Canada, Italy, and Finland.[13] On the other hand, France and the Netherlands are low. Income inequality may account partially for religiosity, but it clearly is not the only factor. If it was, all the countries would be closer to the line.

Fig. 10.1. Relationship between a nation's religiosity and its wealth. Reprinted by permission of the Pew Global Attitudes Project.

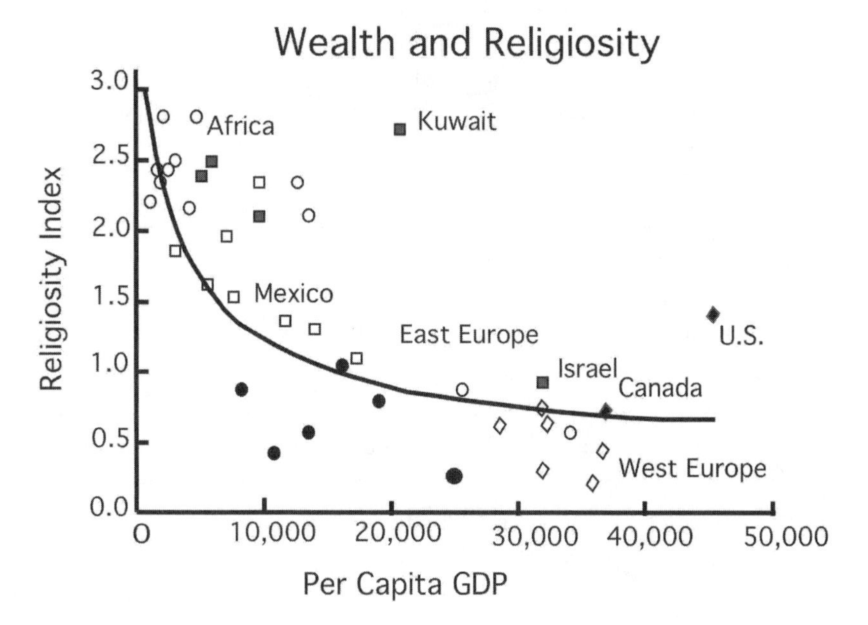

In fact, Norris and Inglehart admit that their data do not reveal an explanation: "In postindustrial nations no empirical support that we have examined could explain why some rich nations are far more religious than others."[14] Or, they might add, some, like France and the Netherlands, less religious.

However, a page later they offer an opinion not supported by their data:

> The United States is exceptionally high in religiosity in large part, we believe, because it is also one of the most unequal postindustrial societies under comparison. Relatively high levels of economic insecurity are experienced by many sectors of U.S. society, despite American affluence, due to the cultural emphasis on the values of personal responsibility, individual achievement and mistrust of big government, limiting the role of public services and the welfare state for basic matters such as health-care covering all the working population. Many American families even in the professional classes, face risks of unemployment, the dangers of

> sudden ill health without private medical insurance, vulnerability to becoming a victim of crime and the problem of paying for long term care for the elderly. Americans face greater anxieties than citizens in other countries about whether they will be covered by medical insurance, whether they will be fired arbitrarily, or whether they will be forced to choose between losing their job and devoting themselves to their new-born child.[15]

As we have known from the time of the philosopher David Hume, correlation does not imply causation. While it may appeal to those who already ideologically view America as dysfunctional to attribute this as the primary source of America's excessive religiosity, the data do not confirm this. While African Americans and blue-collar workers tend to be religious, the white evangelicals you find in megachurch pews and working to get Republicans elected are often very well off. Much of the money that supports the many organizations in the United States that promote extreme conservative views comes from wealthy Christians.

Economic and social inequality probably plays some role in excessive religiosity, for some sectors of American life, but other factors undoubtedly also come in, particularly history. Anyone familiar with the history of these nations should not be surprised to find the United States, Ireland, and Italy on the high side of religiosity, while France sits on the low side.

GODLESS SOCIETY

Religious extremists in America have tried to argue that atheism and secularism will destroy the foundations of society. Televangelist Pat Robertson has asserted that when a society is without religion, "the result will be tyranny."[16] In her best seller *Godless* (same title as the book by atheist Dan Barker), conservative writer Ann Coulter says societies that fail to grasp God's significance are headed toward slavery, genocide, and bestiality.[17] She also asserts that when evolution is widely accepted in a given society, all morality is abandoned.[18] Influential television commentator Bill O'Reilly has said that a society that fails to live "under God" will be a society of "anarchy and crime,"[19] where "lawbreakers are allowed to run wild."[20]

British cleric and theologian Keith Ward has argued that societies lacking strong religious beliefs are immoral, unfree, and irrational.[21] Philosopher John D. Caputo declared that people who are without religion and who do not love God are nothing more than selfish louts, thereby implying that a society with a preponderance of godless people would be a fairly loveless, miserable place.[22]

Once again we see how believers ignore the evidence and make up facts to suit their own prejudices. That's the way faith operates and that's why it should be challenged. Any number of societies now exist where the majority has freely abandoned religion and God. Far from being dens of iniquity, these societies are the happiest, safest, and most successful in the world.

Sociologist Phil Zuckerman spent fourteen months during 2005–2006 in Denmark and Sweden interviewing a wide range of people about their religious beliefs. He presented his results in a 2008 book, *Society without God*.[23]

The first thing Zuckerman noticed upon arriving in Aarhus, Denmark, was: no cops. He saw no police cars, no officers patrolling on foot or motorcycle. A grand total of thirty-one days passed before seeing his first police presence. In 2004 the total number of murders in Aarhus, a bustling metropolis of a quarter million people, was one.

Most Danes and Swedes do not even believe in the notion of "sin," yet their violent crime rates are the lowest in the world. Almost nobody goes to church or reads the Bible. Are they unhappy? In a survey of happiness in ninety-one nations, Denmark ranked number one.[24]

Zuckerman notes that while Denmark and Sweden are societies without God, they are not societies without religion. The majority still pay 1 percent of their annual incomes to support their national churches, get married in church, and baptize their children in church. Most still think of themselves as "Christians." This is done not from faith or spiritual conviction, but out of a sense of cultural tradition. It is unthinkable to most couples not to have a "real wedding" with a "white dress in an old church."[25]

I have observed a similar phenomenon among many of the Jews I have met over the years. Although few admitted it outright, I got the impression that many did not believe in God but nevertheless practiced the rituals of Judaism out of respect for and desire to participate in their ancient heritage. Perhaps this is what we can look forward to as belief declines in America, if it is not already happening—a continued use of the church as a place to celebrate rites of passage, participate in a community, and perhaps even provide a pleasant, quiet place to meditate.

Certainly, one of the appeals of churchgoing is the social life that comes with it, for young and old. Church members are particularly supportive of one another in time of need, such as a death in the family. When I was a graduate student at the University of California at Los Angeles (UCLA), I regularly attended the Single Young Adults group at Westwood Methodist Church, which met every Sunday evening with a program that included a short worship service, a speaker (rarely on a religious topic), and dancing. I met my wife there. I can't see why this cannot continue in a nonsupernatural environment. I am gratified to see atheist and humanist groups that include considerable socializing forming in increasing numbers around America.

Returning to Scandinavia, I need not list all the statistics on Denmark and Sweden to show that belief is very low. Basically about 20 percent believe in a personal God, consider God to be important in their lives, and believe in life after death, compared to about 80 percent in America, depending on how you want to count the effective deists.

What about the health of these godless societies? By every measure of societal health—life expectancy, literacy rates, school enrollment rates, standard of living, infant mortality, child welfare, economic equality, economic competitiveness, gender equality, healthcare, lack of corruption, environmental protection, charity to poor nations, crime, suicide, unemployment—Denmark and Sweden rank near the top.

I don't want to leave the impression that these nations have no problems. Certainly they do, but they are freer to consider rational solutions by not having a majority of citizens who rely on ancient fables for primary guidance.

However, the movement toward increasing secularity in Europe may not continue indefinitely. Eric Kaufmann, a reader in politics and sociology at the University of London, has used birthrate data to project that the current trend on decreasing religiosity in Britain, France, and Scandinavia will continue until around 2050, and then turn around, returning to current levels by 2100.[26] It seems that the future of atheism will depend on two factors: education and economics. If problems with global warming, overpopulation, and energy are not dealt with and even the current rich countries become poor and uneducated, atheism will be one of the first things to go. But, if we can solve these problems, and people everywhere gain a better life, then they will have no further need to hope that some invisible force will someday appear in the sky and put an end to their suffering.

LIVING WITHOUT RELIGION

Several books have come out in recent years in which atheists not directly associated with the new atheist movement have provided their suggestions on how to live without the supposed comfort of religion. Eric Maisel is a PhD whose publisher describes him as a "psychotherapist, philosopher, cultural observer," and "America's foremost creativity coach." He has just released *The Atheist Way: Living Well without Gods*. As Maisel describes it:

> Living the atheist's way is more than living without gods, religion, and supernatural enthusiasms—much more. It is a way of life that integrates the secular, humanist, scientific, freethinking, skeptical, rationalist, and existential traditions into a complete worldview and that rallies that worldview under the banner of atheism, choosing that precise word as its rallying cry. It chooses *atheism* to make clear that our best chance for survival is for members of our species to grow into a mature view of self-interest, one in which human beings can discuss their conflicting interests without one side betraying the other by playing the god card.[27]

The Atheist's Way centers on the question of *meaning*. Instead of assuming that supernatural forces outside ourselves determine our meaning in life, we make our own meaning. Meaning, value, and purpose are human ideas. As Dan Barker puts it, "Instead of a purpose-driven life we have a life-driven purpose." Still, there is something about the book that strikes me as little more than psychobabble self-help with a heavy emphasis on self.

Reference to megachurch pastor Rick Warren's inspirational book of the decade, *The Purpose-Driven Life*,[28] is also made by atheist theologian (yes, there are a few) Robert M. Price in his 2006 book, *The Reason-Driven Life*.[29] In addition to debunking Warren's theology as having no basis in scripture, Price shows that Warren fails to provide a solution to genuine spiritual hunger. Instead, Warren's strict fundamentalist teaching stunts an individual's growth personally, morally, and intellectually. If there is any place where the method of reason promoted by atheism shows its superiority over unquestioned, unsupported faith in helping people live a rewarding life, it is here.

I have just received a copy of *Disbelief 101: A Young Peron's Guide to Atheism* by S. C. Hitchcock (a pseudonym), which I highly recommend.[30]

Atheism is growing most rapidly in just the place it needs to grow, with young people. Atheist groups are multiplying rapidly on campuses and I have been pleased and gratified by the enthusiasm I have found in the colleges and universities I have visited over the past two years.

Finally, let me mention *Living without God* by the distinguished atheist philosopher Ronald Aronson.[31]

Aronson is critical of the new atheists in not providing secularists with an alternative to believe in if they are not to believe in God. They only provide denial of God and nothing to compare with what he calls the "coherence" of religious belief:

> Our religious friends affirm their belief in the coherence of the universe and the world, their deep sense of belonging to it and to a human community, their refusal to be stymied by the limits of knowledge, their confidence in dealing with life's mysteries and uncertainties, their willingness to take complete responsibility for the small things while leaving forces beyond themselves in charge of the large ones, their security in knowing right from wrong, and perhaps above all, their sense of hope about the future.[32]

Aronson then adds,

> Even if we would reject these beliefs as unfounded and irrational, we have to be struck by their force. And envy their coherence.... Besides disbelief, what do we have to offer? What should we tell our children and grandchildren as we see them swept up in a pervasively religious environment.[33]

The new atheists do indeed reject religious beliefs as irrational and are certainly not going to dream up other irrational beliefs to take their place. We strongly disagree with Aronson that they are "coherent." It is not coherent to kill for your religious beliefs. The existential questions—death, loss, suffering, and inhumanity—are not answered by the great religions, unless you think "it's a mystery" is an answer.

We also disagree with Aronson's implication that holding to a set of irrational beliefs can be healthy. And we have plenty to offer besides disbelief. We have freedom of thought and the ability to live our lives the way we want to live them without anybody forcing superstitious rules upon us.

As for what to tell our children and grandchildren, we tell them nothing! We teach them to think for themselves and then trust them to arrive at rational conclusions that suit themselves and serve them in their lives, which belong to them and not to us. My wife and I hardly ever talked about religion at the dinner table and even sent our children to church-connected private schools, at great financial sacrifice, where they would get the best possible education. One of the schools was Punahou in Hawaii, where Barack Obama graduated a few years ahead of our daughter. Our kids attended chapel and never heard us tell them what to believe. Long before I started writing books on the subject they had decided for themselves that there is no God.

Now, Aronson is an atheist and claims he will go beyond the New Atheism and tell the rest of us how to live without God. He starts by pointing out that hundreds of years after the Enlightenment it is still very hard to be human, with death, loss, suffering, and inhumanity seemingly inescapable parts of life. Furthermore, "despite" our technological achievements we still are ruining the environment, insecure about healthcare, vulnerable economically, and have an unequal and unfair distribution of wealth. Theists use religion to ease these difficulties of living, and living without God requires finding nontheistic relief from these same difficulties.

Although this is a fine statement of our problem, Aronson does not come close to providing sufficient answers. He presents some bland essays on "gratitude and dependence," "aging and dying," and "responsibility and hope," but much more is needed. Aronson usefully provides some of the thoughts of classical thinkers about godlessness, how Albert Camus decided life must always be absurd while John-Paul Sartre "looked beyond absurdity to human-imposed order, meaning, and purpose."[34] Clearly Maisel (see above) was not first with this idea.

New Atheism is, well, new. By the time it becomes old we will surely have more answers. But I submit that it has broken some new ground and already provides some viable ideas about how to live in a godless universe. At this point, let me mention the works of a few recent atheist authors whom I have not discussed previously in this book but who have added considerably to the intellectual case for nonbelief: *Atheist Universe: Why God Didn't Have a Thing to Do with It* by self-help author David Mills;[35] *50 Reasons People Give for Believing in a God* by Guy P. Harrison;[36] *Atheism Explained: From Folly to Philosophy* by sociologist David Ramsay Steele;[37] *Encountering*

Naturalism: A Worldview and Its Uses by the director of the Center for Naturalism, Thomas W. Clark;[38] *The Age of American Unreason* by Susan Jacoby;[39] *The Secular Conscience: Why Belief Belongs in Public Life* by Austin Dacey;[40] and *The Ghost in the Universe: God in Light of Modern Science*[41] by physicist Taner Edis. I highly recommend these to the reader for further study, although I do not claim these authors endorse all the views I have presented or consider themselves "new atheists."

THE NEW ATHEISM SUMMARIZED

In this book I have reviewed and expanded on the ideas that have become associated with New Atheism as presented in the best-selling books that appeared between 2004 and 2009 by Sam Harris, Richard Dawkins, Daniel C. Dennett, Christopher Hitchens, and myself. We have departed from the views of traditional atheists in several respects, and, as we have seen, some added their voices to those of theists to object.

The new atheists' philosophy is summarized well in the titles and subtitles of their books. Harris urges the "end of faith" since it is worthless and dangerous to believe without evidence. Dawkins sees belief in God as a "delusion." Dennett urges us to "break the spell" of the taboos about studying and questioning religion. Hitchens writes about "how religion poisons everything." And I claim that not only is there no evidence for God, "science shows that God does not exist."

Some atheists criticize these positions as too uncompromising. They say that we need to work together with "moderate" religious groups if we are to keep creationism out of the schools. New atheists do not want to see creationism in the schools either, but do not regard it as such an overarching problem that we can ignore the even greater damage that is done to society by the irrational thinking associated with religion. As long as moderate believers continue to promote an unsupported faith that claims divine revelation as a source of knowledge, they encourage the extreme elements of that faith to feel free to commit any horrific act, thinking they are carrying out the will of God.

Theistic critics accuse New Atheism of "scientism," which is the principle that science is the *only* means that can be used to learn about the world and humanity. They cannot quote a single new atheist who has said that. We fully recognize the value of and participate in other realms of thought and activity such as art, music, literature, poetry, and moral philosophy. At the same time, where observed phenomena are at issue, we insist that scientific method has a proper role. This includes questions of the supernatural and the existence of any god who actively engages in the affairs of the universe.

Theists also try to argue that science operates on faith no less than does religion by assuming science and reason apply to reality. This betrays an ignorance of science that is pervasive among theists and theologians. Faith is belief in the absence of evidence. Science is belief in the presence of evidence. When the evidence disagrees with a scientific proposition, the proposition is discarded. When the evidence disagrees with a religious proposition, the evidence is discarded.

Theists similarly misunderstand the use of reason. They say you can't prove the universe is reasonable without making a circular argument, assuming what it is you are trying to prove. If so, their argument is circular too, because they are using reason. But this is not how it works. It is not the universe that is reasonable or not. It is people who are reasonable or not. Reason and logic are just ways of thinking and speaking that are designed to ensure that a concept is consistent with itself and with the data. How can you expect to learn anything from inconsistent, irrational thinking and speaking?

In this regard, I wish to emphasize that our dispute with believers is purely an intellectual one. There is no reason why it cannot be carried on civilly. We are often criticized for engaging in "polemics" against religion, but my dictionary defines polemic as "a passionate argument." What's wrong with being passionate about one's arguments? Believers are equally passionate about theirs. We see nothing wrong with asking the faithful to provide evidence and rational arguments for their faith, rather than us keeping quiet as do so many of our atheist and agnostic friends for fear of offending "deeply held beliefs." And why should we be faulted for bringing up the historical facts about the atrocities committed in the name of God? We have every right as scholars to point out the inconsistencies and downright errors in scriptures, and to describe the historical and archaeological finds that prove beyond a reasonable doubt that events like Exodus never happened and that the Jesus described in the New Testament is largely if not wholly a mythological figure. We are not trying to shut down religious institutions, but we have a right as citizens to object to the illegal and unconstitutional distribution of taxpayer money to these institutions and other special privileges they are awarded by cowardly politicians. And, we feel it is our duty to protest when government officials rely on superstition instead of science to make decisions that affect the lives of everyone on this planet.

I have expanded on the scientific position of New Atheism, which holds that a completely materialistic model of the universe provides a plausible explanation for all our observations, from cosmology to the human mind, leaving no gap for God or the supernatural to be inserted.

Theists think they have a gap for God provided by the big bang that requires a supernatural creation. I have shown that this claim is based on the mistaken notion that the universe must have had a beginning of infinite density and infinitesimal size called a "singularity." They wrongly interpret this singularity as the beginning of space and time.

Stephen Hawking was one of the authors of the original "proof" three decades ago that such a singularity follows from Einstein's theory of general relativity. Two decades ago, in his best seller *A Brief History of Time*, Hawking explained that the original calculation did not take into account quantum mechanics and that, when this is done, the singularity does not occur.

Even in their most recent books and lectures, theists continue to ignore Hawking and propagate this error. They don't seem to care since they believe their audiences are generally too unsophisticated to pick it up. The universe probably had no beginning. In one plausible scenario that can be completely worked out mathematically, our universe appeared from an earlier one that always existed by a well-understood process called quantum tunneling.

The same misunderstanding of or disregard for the scientific facts holds for another theist claim, that the universe is so fine-tuned for life that it could only have been created by an intelligent being. Even my most prominent new atheist colleagues have trouble with this one because they are not physicists. In this book, and earlier ones, I have tried to explain why the fine-tuning argument fails. The parameters of nature are not fine-tuned at all. Either they have the values they do because of being arbitrary to begin with, being fixed by the laws of physics, or else they still allow for *some* kind of life when other parameters are varied. A common mistake by the unqualified people making these claims of fine-tuning is to fix all the parameters but one and then vary just that one. Any properly trained scientist knows that all parameters must be varied when studying what happens when the parameters of a system change.

The intelligent design movement in biology is dying a natural death, although we will probably hear a few death throes before final interment. Intelligent design in cosmology deserves to die. I am tired of shooting so many arrows into it. The new area in science where theists are beginning to stake out a position is the mind. They think they can find a gap for God in the fact that neuroscience still does not have a consensus on a purely material model for the operations of the brain that are associated with thinking and consciousness. However, we have seen that all the evidence points to the mind as a purely material phenomenon and the only valid objections to this conclusion are theological or philosophical. Claims of evidence for special powers of the mind, such as ESP, do not stand up under the application of the same strict criteria used in assessing any extraordinary claim in science.

I have given examples of the atrocities in the Bible and those committed by religion over the ages. I spent some time on the Mormons since they represent a sect that formed sufficiently recently that we have reliable historical data on which to draw some conclusions about why religion and violence go together like a horse and carriage. I like simple answers, and the answer is simple. A person is capable of slitting a baby's throat if he is convinced he is following God's orders.

I also have countered some of the criticism of the new atheists, particularly aimed at Sam Harris and Christopher Hitchens, who blame September 11 and other terrorist acts squarely on religion and not the sole result of political oppression. As we saw in chapter 5, the last instructions Mohamed Atta gave to his team make this perfectly clear. Islam flew those planes into those buildings.

Many of the theist books critical of New Atheism attempt to argue that "atheists" like Hitler and Stalin killed more people than all the kings of Christendom, the Crusades, and the Inquisition put together. This assertion is invalid on several fronts. First, there is no atheist scripture that tells people to kill people in the name of atheism, the way the Bible and the Qur'an condone the killing of nonbelievers. Second, Hitler was not an atheist. Third, if there were no religion there would not be a separate, hated group of people like the Jews who over the centuries have been persecuted for "killing Christ." Fourth, recently released documents from the Soviet era prove that Stalin did not kill in the name of atheism and in fact normalized relations with the Russian Orthodox Church during World War II. Churches everywhere have often worked closely with governments no matter how cruel to suppress opposition. The Catholic Church aided the Nazis; the Russian Orthodox Church helped keep the czars in power; both the Soviet and today's Russian rulers continue to find it useful in this regard.

One does not measure evil by the numbers killed. Is killing ten innocent people worse than killing one? Every king, pope, crusader, or inquisitor who unjustly killed a single person committed an evil act. And history records far more people who killed in the name of religion than who did so in the name of unbelief.

I spent some time summarizing the conclusions of biblical scholar Bart Ehrman, who after serving as a Protestant minister lost his faith when he found that the Bible and Christian teaching in general fail to account for the suffering in the world. None of the biblical answers work. The prophets taught that suffering was God's punishment for disobedience. But even the obedient suffer, as the story of Job shows. The claim that suffering is caused by humans, made possible by free will, fails to explain suffering from natural disasters. Another idea is that suffering is redemptive, bringing good where there was previously evil. But what possible good can come of the many children in the world who starve to death each day? Finally, the Bible proposes that suffering and death will end when Jesus returns for the Last Judgment. Jesus promised it would happen in a generation. It hasn't yet.

I briefly reviewed the answers to suffering provided by other religions besides Christianity. In Islam, it's simply God's will not to be questioned. Hindus see suffering alleviated in a future life. Buddha taught an eightfold path to end one's suffering, which comes with the elimination of the ego when one's sequence of lives terminates in nothingness. Taoism also preaches an end to the love of self and its replacement with a love of the world.

Sam Harris suggested in *The End of Faith* that we pay attention to some of the insights of Buddha and other spiritualists and mystics of the East to aid us in understanding our own minds. Religious historian Karen Armstrong has written about the period from about 900 BCE to 200 BCE, peaking around 500 BCE, when a great transformation occurred in human thinking. An equivalent revolution in human thinking was not repeated again until 1600–1700 CE with the rise of science and reason in the West. This earlier period is called the Axial Age and is characterized by thinkers in India and China turning inward and seeking to find truths within themselves by emptying their minds and freeing themselves of the dominance of ego. Armstrong also describes transformations in thinking taking place in Greece and Israel, but these led in a different direction.

Remarkably, the primary sages Buddha, Lao Tzu, and Confucius all lived around 500 BCE. Furthermore, Thales of Miletus, the first scientist in recorded history, supposedly predicted the eclipse of the sun that occurred May 29, 586 BCE, triggering the Greek axial age. Although this event is disputed, Greek science nevertheless began with Thales. Coincidentally, the Jews were hauled off to exile in Babylonia in 585 BCE, leading to the axial age in Judaism.

While a handful of mystics in the Abrahamic religions also taught the virtues of selflessness, the great bulk of adherents to these faiths were drawn in by highly self-centered thinking. This is perhaps less so in Judaism, but both Christianity and Islam are the two most popular religions today for one reason more than any other: the promise of eternal life.

Atheism can never compete with this promise. We will not win a single convert by promising nothingness in the place of paradise. (Hell is only promised for others.) However, if you are already an atheist and view paradise as the impossible dream that it is, then you need a way to face nothingness. The great sages have pointed the way, and if we eliminate the superstitions and supernatural detritus that the religions they founded have accumulated over the last 2,500 years, we have what I have termed the Way of Nature. We are natural, material beings with natural, material brains that have, by whatever means, evolved to give us an exalted sense of self.

Theists assert that religion is necessary for a moral society, that morality comes from God, and that in an atheist society everyone would run wild. This is certainly not born out by the facts, which indicate that problems with crime, child and spousal abuse, teen pregnancy, drug abuse, and other social ills are lower in the less religious nations. The evidence is overwhelming that the happiest, best-adjusted, healthiest societies in the world are those in which the majority has freely abandoned belief in God.

While scholars are still debating mechanisms, it seems highly likely that morality evolved naturally in society and was only adopted later, imperfectly, by religion. Many biblical teachings, such as support for slavery and subjugation of women, are immoral by modern standards. These examples also show how morality is not constant but evolves with time.

And what specific values to live by are suggested for your consideration by New Atheism? They are no different than those associated with humanism and general philosophical rationality. They can be found in a wealth of literature, especially the many works of Bertrand Russell. More recent summaries include *The Philosophy of Humanism* by Corliss Lamont,[42] *Eupraxophy* by Paul Kurtz,[43] *How to Live* by Peter Singer,[44] and *Value and Virtue in a Godless Universe* by Erik J. Wielenberg.[45] The last author offers this thought:

> The victory of accepting naturalism consists in exerting control over one's own mind. In a universe in which human beings are largely at the mercy of morally indifferent forces beyond their control, one prominent kind of achievement is *taking control*. In the particular case of accepting naturalism, the victory is over fear. The religious believer is driven by fear; in this way, the believer's mind is subject to forces beyond personal control just as much as the body. But the naturalist takes control of the mind and refuses to be ruled by fear; this victory over the universe is a worthwhile achievement in and of itself.[46]

The message of New Atheism has been terribly misunderstood as being exclusively negative. For every negative we have an even greater positive. Faith is absurd and dangerous and we look forward to the day, no matter how distant, when the human race finally abandons it. Reason is a noble substitute, proven by its success. Religion is an intellectual and moral sickness that cannot endure forever if we believe at all in human progress. Science sees no limit in the human capacity to comprehend the universe and ourselves. God does not exist. Life without God means we are the governors of our own destinies.

If you are a theist or other believer, throw off your yoke and join us. If you are an agnostic, look at the evidence and see that we do in fact know that God does not exist and join us. If you are an atheist who thinks we should work with moderate believers, look at the consequences of irrational thought and join us.

NOTES

1. Dan Barker, *Godless: How an Evangelical Preacher Became One of America's Leading Atheists* (Berkeley, CA: Ulysses Press, 2008), p. 40.

2. Dean H. Hamer, *The God Gene: How Faith Is Hardwired into Our Genes* (New York: Doubleday, 2004); Eugene G. D'Aquili and Andrew B. Newberg, *The Mystical Mind: Probing the Biology of Religious Experience* (Minneapolis, MN: Fortress Press, 1999).

3. Scott Atran, *In Gods We Trust: The Evolutionary Landscape of Religion* (Oxford: Oxford University Press, 2002), p. 186ff.

4. Hector Avalos, *Fighting Words: The Origins of Religious Violence* (Amherst, NY: Prometheus Books, 2005), p. 330.

5. Nicholas Wade, *Before the Dawn: Recovering the Lost History of Our Ancestors* (New York: Penguin, 2006), p. 30.

6. Alexander Saxton, "The Great God Debate and the Future of Faith," *Free Inquiry* 29, no. 1 (2009): 37–43. Original quotation in Victor J. Stenger, *God: The Failed Hypothesis* (Amherst, NY: Prometheus Books, 2007), pp. 245–46.

7. Ibid., p. 43.

8. Hank Davis, *Caveman Logic: The Persistence of Primitive Thinking in a Modern World* (Amherst, NY: Prometheus Books, 2009).

9. Pippa Norris and Ronald Inglehart, *Sacred and Secular: Religion and Politics Worldwide*, Cambridge Studies in Social Theory, Religion, and Politics (Cambridge: Cambridge University Press, 2004).

10. Ibid., p. 3.

11. Rodney Stark and Roger Finke, *Acts of Faith: Explaining the Human Side of Religion* (Berkeley: University of California Press, 2000), p. 79.

12. Gregory Paul and Phil Zuckerman, "Why the Gods Are Not Winning," *Edge: The Third Culture*, http://www.edge.org/3rd_culture/paul07/paul07_index.html (accessed February 8, 2009).

13. Norris and Inglehart, *Sacred and Secular*, p. 107.

14. Ibid., p. 106.

15. Ibid., pp. 107–108.

16. As quoted from the film *With God on Our Side: George W. Bush and the Rise of the Religious Right*, by Calvin Skaggs, David Van Taylor, and Ali Pomeroy, Lumiere Productions, 2004.

17. Ann H. Coulter, *Godless: The Church of Liberalism* (New York: Crown Forum, 2006), p. 3. No to be confused with Dan Barker's *Godless*.

18. Ibid., p. 280.

19. Bill O'Reilly, *Culture Warrior* (New York: Broadway Books, 2006), p. 19.

20. Ibid., p. 72.

21. Keith Ward, *In Defence of the Soul* (Oxford: Oneworld, 1998).

22. John D. Caputo, *On Religion* (New York: Routledge, 2001).

23. Phil Zuckerman, *Society without God: What the Least Religious Nations Can Tell Us about Contentment* (New York: New York University Press, 2008).

24. "Something's Happy in the State of Denmark," *Los Angeles Times*, June 19, 2006.

25. Zuckerman, *Society without God*, p. 9.

26. Eric Kaufmann, "Sacralization by Stealth: Demography, Religion, and Politics in Europe," *Jewish Policy Research*, http://www.jpr.org.uk/downloads/Kaufmann%20Paper.pdf (accessed February 25, 2009).

27. Eric Maisel, *The Atheist's Way: Living Well without Gods* (Novato, CA: New World Library, 2009).

28. Richard Warren, *The Purpose-Driven Life: What on Earth Am I Here For?* (Grand Rapids, MI: Zondervan, 2002).

29. Robert M. Price, *The Reason-Driven Life: What Am I Here on Earth For?* (Amherst, NY: Prometheus Books, 2006).

30. S. C. Hitchcock, *Disbelief 101: A Young Person's Guide to Atheism* (Tucson, AZ: Sharp Press, 2009).

31. Ronald Aronson, *Living without God: New Directions for Atheists, Agnostics, Secularists, and the Undecided* (Berkeley, CA: Counterpoint, 2008).

32. Ibid., p. 17.

33. Ibid.

34. Ibid., p. 45.

35. David Mills, *Atheist Universe: Why God Didn't Have a Thing to Do with It* (Philadelphia: Xlibris, 2003).

36. Guy P. Harrison, *50 Reasons People Give for Believing in a God* (Amherst, NY: Prometheus Books, 2008).

37. David Ramsay Steele, *Atheism Explained: From Folly to Philosophy* (Chicago: Open Court, 2008).

38. Thomas W. Clark, *Encountering Naturalism: A Worldview and Its Uses* (Somerville, MA: Center for Naturalism, 2007).

39. Susan Jacoby, *The Age of American Unreason* (New York: Pantheon Books, 2008).

40. Austin Dacey, *The Secular Conscience: Why Belief Belongs in Public Life* (Amherst, NY: Prometheus Books, 2008).

41. Taner Edis, *The Ghost in the Universe: God in Light of Modern Science* (Amherst, NY: Prometheus Books, 2002).

42. Corliss Lamont, *The Philosophy of Humanism*, 8th ed. (Amherst, NY: Humanist Press, 1996). First published in 1965.

43. Paul Kurtz, *Eupraxophy: Living without Religion* (Amherst, NY: Prometheus Books, 1989).

44. Peter Singer, *How Are We to Live? Ethics in an Age of Self-Interest* (Oxford: Oxford University Press, 1997).

45. Erik J. Wielenberg, *Value and Virtue in a Godless Universe* (Cambridge: Cambridge University Press, 2005).

46. Ibid., p. 118.

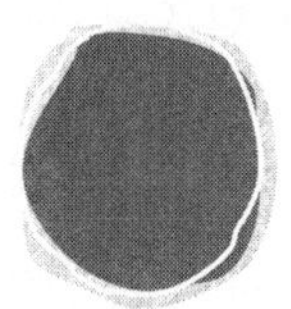

BIBLIOGRAPHY

Aikman, David. *The Delusion of Disbelief: Why the New Atheism Is a Threat to Your Life, Liberty, and Pursuit of Happiness*. Carol Stream, IL: Tyndale House, 2008.

Armstrong, Karen. *The Great Transformation: The Beginning of Our Religious Traditions*. New York: Anchor, 2007.

Arntz, William, Betsy Chasse, and Mark Vicente. *What the Bleep Do We Know!? Discovering the Endless Possibilities for Altering Your Everyday Reality*. Deerfield Beach, FL: Health Communications, 2005.

Aronson, Ronald. *Living without God: New Directions for Atheists, Agnostics, Secularists, and the Undecided*. Berkeley, CA: Counterpoint, 2008.

Atkatz, David. "Quantum Cosmology for Pedestrians." *American Journal of Physics* 62 (1994): 619–27.

Atran, Scott. *In Gods We Trust: The Evolutionary Landscape of Religion*. Oxford: Oxford University Press, 2002.

Augustine, Keith. "Hallucinatory Near-Death Experiences." http://www.infidels.org/library/modern/keith_augustine/HNDEs.html#experiments (accessed December 11, 2008).

Avalos, Hector. *Fighting Words: The Origins of Religious Violence*. Amherst, NY: Prometheus Books, 2005.

Azkoul, Michael. *Anti-Christianity, the New Atheism*. Montreal: Monastery Press, 1984.

Baars, Bernard J. "The Conscious Access Hypothesis: Origins and Recent Evidence." *TRENDS in Cognitive Sciences* 6, no. 1 (2002): 47–62.

Barker, Dan. *Godless: How an Evangelical Preacher Became One of America's Leading Atheists.* Berkeley, CA: Ulysses Press, 2008.

Barrau, Aurélian. "Physics in the Multiverse." http://physicaplus.org.il/zope/home/en/1223032001/barrau_en/ (accessed June 5, 2009).

Barrow, John D., and Frank J. Tipler. *The Anthropic Cosmological Principle.* Oxford: Oxford University Press, 1988.

Bartholomew, David J. *God, Chance, and Purpose: Can God Have It Both Ways?* Cambridge: Cambridge University Press, 2008.

Beattie, Tina. *The New Atheists: The Twilight of Reason and the War on Religion.* Maryknoll, NY: Orbis Books, 2007.

Beauregard, M., and V. Paquette. "Neural Correlates of a Mystical Experience in Carmelite Nuns." *Neuroscience Letters* 405 (2006): 186–90.

Beauregard, M., V. Paquette, and J. Levesque. "Dysfunction in the Neural Circuitry of Emotional Self-Regulation in Major Depressive Disorder." *Neuroreport* 17 (2006): 843–46.

Behe, Michael J. *Darwin's Black Box: The Biochemical Challenge to Evolution.* New York: Free Press, 1996.

Benedict, Stephan Otto Horn, Siegfried Wiedenhofer, and Christoph von Schönborn. *Creation and Evolution: A Conference with Pope Benedict XVI in Castel Gandolfo.* San Francisco: Ignatius Press, 2008.

Benson, H., J. A. Dusek, J. B. Sherwood, P. Lam, C. F. Bethea, et al. "Study of the Therapeutic Effects of Intercessory Prayer (Step) in Cardiac Bypass Patients: A Multicenter Randomized Trial of Uncertainty and Certainty of Receiving Intercessory Prayer." *American Heart Journal* 154, no. 4 (2007): 934–42.

Berlinski, David. *The Devil's Delusion: Atheism and Its Scientific Pretensions.* New York: Crown Forum, 2008.

Beversluis, John. *C. S. Lewis and the Search for Rational Religion.* Amherst, NY: Prometheus Books, 2007.

Blackmore, Susan J. *Dying to Live: Near-Death Experiences.* Amherst, NY: Prometheus Books, 1993.

Blaker, Kimberley. *The Fundamentals of Extremism: The Christian Right in America.* New Boston, MI: New Boston Books, 2003.

Boston, Rob. "Faith-Based Flare Up." *Church & State* 61, no. 8 (2008): 4–6.

Boyer, Pascal. *Religion Explained: The Evolutionary Origins of Religious Thought.* New York: Basic Books, 2001.

Brockman, John. "Edge: The Reality Club." http://www.edge.org/3rd_culture/coyne09/coyne09_index.html (accessed April 8, 2009).

Broom, Donald M. *The Evolution of Morality and Religion.* Cambridge: Cambridge University Press, 2003.

Byers, Nina. "E. Noether's Discovery of the Deep Connection between Symmetries and Conservation Laws." *Israel Mathematical Conference.* http://www.physics.ucla.edu/~cwp/articles/noether.asg/noether.html (accessed February 20, 2009).

Byrne, Rhonda. *The Secret.* New York: Attria Books, 2006.

Capra, Fritjof. *The Tao of Physics: An Exploration of the Parallels between Modern Physics and Eastern Mysticism.* London: Wildwood House, 1975.

Caputo, John D. *On Religion.* New York: Routledge, 2001.

Carrier, Richard C. "Defending Naturalism as a Worldview: A Rebuttal to Michael Rea's *World without Design.*" http://www.infidels.org/library/modern/richard_carrier/rea.html#1, 2003 (accessed June 5, 2009).

Chomsky, Noam. *9-11.* New York: Open Media/Seven Stories Press, 2001.

Chopra, Deepak. *Ageless Body, Timeless Mind: The Quantum Alternative to Growing Old.* New York: Harmony Books, 1993.

———. *Quantum Healing: Exploring the Frontiers of Mind/Body Medicine.* New York: Bantam Books, 1989.

Churchland, Patricia Smith. *Neurophilosophy: Toward a Unified Science of the Mind-Brain.* Cambridge, MA: MIT Press, 1995.

Churchland, Paul M. *The Engine of Reason, the Seat of the Soul: A Philosophical Journey into the Brain.* Cambridge, MA: MIT Press, 1995.

Clark, Thomas W. *Encountering Naturalism: A Worldview and Its Uses.* Somerville, MA: Center for Naturalism, 2007.

Collins, Francis S. *The Language of God: A Scientist Presents Evidence for Belief.* New York: Free Press, 2006.

Cornwell, John. *Darwin's Angel: A Seraphic Response to* The God Delusion. London: Profile Books, 2007.

Coulter, Ann H. *Godless: The Church of Liberalism.* New York: Crown Forum, 2006.

Coyne, Jerry A. "Seeing and Believing: The Never-Ending Attempt to Reconcile Science and Religion, and Why It Is Doomed to Fail." *New Republic*, February 4, 2009.

Crean, Thomas. *God Is No Delusion: A Refutation of Richard Dawkins.* San Francisco: Ignatius Press, 2007.

Dacey, Austin. *The Secular Conscience: Why Belief Belongs in Public Life.* Amherst, NY: Prometheus Books, 2008.

Dalai Lama, and Howard C. Cutler. *The Art of Happiness: A Handbook for Living.* New York: Riverhead Books, 1998.

D'Aquili, Eugene G., and Andrew B. Newberg. *The Mystical Mind: Probing the Biology of Religious Experience.* Minneapolis: Fortress Press, 1999.

Davies, Paul. "Taking Science on Faith." *New York Times*, November 24, 2007.

Davis, Hank. *Caveman Logic: The Persistence of Primitive Thinking in a Modern World.* Amherst, NY: Prometheus Books, 2009.

Dawkins, Richard. *The Ancestor's Tale: A Pilgrimage to the Dawn of Evolution.* Boston: Houghton Mifflin, 2004.

———. *The Blind Watchmaker: Why the Evidence of Evolution Reveals a Universe without Design.* New York: Norton, 1996.

———. *Climbing Mount Improbable.* 1st American ed. New York: Norton, 1996.

———. *The Extended Phenotype: The Gene as the Unit of Selection.* Oxford: Freeman, 1982.

———. *The God Delusion.* Boston: Houghton Mifflin, 2006.

———. *The Selfish Gene.* New York: Oxford University Press, 1976.

———. *River out of Eden: A Darwinian View of Life.* Science Masters series. New York: Basic Books, 1995.

———. *Unweaving the Rainbow: Science, Delusion, and the Appetite for Wonder.* Boston: Houghton Mifflin, 1998.

Dawkins, Richard, and Daniel Clement Dennett. *The Extended Phenotype: The Long Reach of the Gene.* Rev. ed. Oxford: Oxford University Press, 1999.

Dawkins, Richard, and Latha Menon. *A Devil's Chaplain: Selected Essays.* London: Weidenfeld & Nicolson, 2003.

Day, Vox. *The Irrational Atheist: Dissecting the Unholy Trinity of Dawkins, Harris, and Hitchens.* Dallas/Chicago: BenBella Books Distributed by Independent Publishers Group, 2008.

Dembski, William A. *Intelligent Design: The Bridge between Science & Theology.* Downers Grove, IL: InterVarsity Press, 1999.

Dennett, Daniel Clement. *Brainchildren: Essays on Designing Minds.* Cambridge, MA: MIT Press, 1998.

———. *Brainstorms: Philosophical Essays on Mind and Psychology.* Montgomery, VT: Bradford Books, 1978.

———. *Breaking the Spell: Religion as a Natural Phenomenon.* New York: Viking, 2006.

———. *Consciousness Explained.* Boston: Little, Brown, 1991.

———. *Content and Consciousness.* 2nd ed. International Library of Philosophy and Scientific Method. London: Routledge & Kegan Paul, 1986.

———. *Darwin's Dangerous Idea: Evolution and the Meanings of Life.* New York: Simon & Schuster, 1995.

———. *Elbow Room: The Varieties of Free Will Worth Wanting.* Oxford: Oxford University Press/New York: Clarendon Press, 1984.

———. *Freedom Evolves.* New York: Viking, 2003.

———. *The Intentional Stance.* Cambridge, MA: MIT Press, 1987.

———. *Kinds of Minds: Toward an Understanding of Consciousness.* New York: Basic Books, 1996.

———. *Sweet Dreams: Philosophical Obstacles to a Science of Consciousness.* Jean Nicod lectures. Cambridge, MA: MIT Press, 2005.

De Rosa, Peter. *Vicars of Christ: The Dark Side of the Papacy.* New York: Crown, 1988.

Doherty, Earl. *The Jesus Puzzle: Did Christianity Begin with a Mythical Christ?* Ottawa: Canadian Humanist Publications, 1999.

Drange, Theodore M. *Nonbelief & Evil: Two Arguments for the Nonexistence of God.* Amherst, NY: Prometheus Books, 1998.

Draper, John William. *History of the Conflict between Religion and Science.* London: Watts, 1873.

Draper, Paul. "God or Blind Nature? Philosophers Debate the Evidence." http://www.infidels.org/library/modern/debates/great-debate.html (accessed November 20, 2008).

———. "Naturalism." http://ndpr.nd.edu/review.cfm?id=14725 (accessed November 18, 2008).

D'Souza, Dinesh. *What's So Great about Christianity?* Washington, DC: Regnery, 2007.

Dyer, Wayne W. *Change Your Thoughts, Change Your Life: Living the Wisdom of the Tao.* Carlsbad, CA: Hay House, 2007.

Edis, Taner. "A False Quest for a True Islam." *Free Inquiry* 25, no. 5 (2007): 48–50.

———. *The Ghost in the Universe: God in Light of Modern Science.* Amherst, NY: Prometheus Books, 2002.

———. "How Gödel's Theorem Supports the Possibility of Machine Intelligence." *Minds and Machines* 8 (1998): 251–62.

———. *An Illusion of Harmony: Science and Religion in Islam.* Amherst, NY: Prometheus Books, 2007.

———. *Science and Nonbelief.* Amherst, NY: Prometheus Books, 2008.

Egnor, Michael, "Ideas, Matter, and Faith." http://www.evolutionnews.org/2007/06/ideas_matter_and_dogma.html (accessed January 12, 2009).

Ehrman, Bart D. *God's Problem: How the Bible Fails to Answer Our Most Important Question—Why We Suffer.* New York: HarperOne, 2008.

———. *Misquoting Jesus: The Story behind Who Changed the Bible and Why.* New York: HarperSanFrancisco, 2005.

Elbert, Jerome. *Are Souls Real?* Amherst, NY: Prometheus Books, 2000.

Farmer, Steve. "Neurobiology, Primitive Gods and Textual Traditions: From Myth to Religions and Philosophies." *Cosmos: Journal of the Traditional Cosmology Society* 22, no. 2 (2006): 55–119.

Feser, Edward. *The Last Superstition: A Refutation of the New Atheism.* South Bend, IN: St. Augustine's Press, 2008.

Finkelstein, Israel, and Neil Asher Silberman. *The Bible Unearthed: Archaeology's New Vision of Ancient Israel and the Origin of Its Sacred Texts.* New York: Free Press, 2001.

Flanagan, Owen J. *The Really Hard Problem: Meaning in a Material World.* Cambridge, MA: MIT Press, 2007.

Flynn, Tom. *The New Encyclopedia of Unbelief.* Amherst, NY: Prometheus Books, 2007.

Forrest, Barbara, and Paul R. Gross. *Creationism's Trojan Horse: The Wedge of Intelligent Design.* Oxford: Oxford University Press, 2004.

Franklin, Stan. *Artificial Minds.* Cambridge, MA: MIT Press, 1995.

Freedman, Samuel G. "For Atheists, Politics Proves to Be a Lonely Endeavor." *New York Times,* October 18, 2008.

Frith, Chris. *Making Up the Mind: How the Brain Creates Our Mental World.* Oxford: Blackwell, 2007.

Froese, Paul. *The Plot to Kill God: Findings from the Soviet Experiment in Secularization.* Berkeley: University of California Press, 2008.

Gamez, David. "Progress in Machine Consciousness." *Consciousness and Cognition* 17 (1008): 887–910.

Garrison, Becky. *The New Atheist Crusaders and Their Unholy Grail: The Misguided Quest to Destroy Your Faith.* Nashville: Thomas Nelson, 2007.

Giberson, Karl. *Saving Darwin: How to Be a Christian and Believe in Evolution.* New York: HarperOne, 2008.

Gödel, Kurt. "On Formally Undecidable Propositions of *Principia Mathematica* and Related Systems." *Monatshefte für Mathematik und Physik* 38 (1931): 173–98.

Goetz, Stewart, and Charles Taliaferro. *Naturalism.* Grand Rapids, MI: William B. Erdman, 2008.

Goldberg, Michelle. *Kingdom Coming: The Rise of Christian Nationalism.* New York: Norton, 2006.

Good, I. J. "Where Has the Billion Trillion Gone?" *Nature* 389, no. 6653 (1997): 806–807.

Gould, Stephen Jay. *Rocks of Ages: Science and Religion in the Fullness of Life.* Library of Contemporary Thought. New York: Ballantine, 1999.

Greene, Susan. "Polygamy Prevails in Remote Arizona Town." *Denver Post,* March 4, 2001.

Gunaratana, Henepola. *Mindfulness in Plain English.* Boston: Wisdom Publications, 2002.

Haggard, P., C. Newman, and E. Magno. "On the Perceived Time of Voluntary Actions." *British Journal of Psychology* 90, no. 2 (1999): 291–303.

Hahn, Scott. *Answering the New Atheism: Dismantling Dawkins' Case against God.* Steubenville, OH: Emmaus Road, 2008.

Hamer, Dean H. *The God Gene: How Faith Is Hardwired into Our Genes.* New York: Doubleday, 2004.

Harris, Melvin. "Are 'Past-Life' Regressions Evidence for Reincarnation?" *Free Inquiry* 6 (1986): 18.

Harris, Sam. *The End of Faith: Religion, Terror, and the Future of Reason.* New York: Norton, 2004.

———. *Letter to a Christian Nation*. New York: Knopf, 2006.

Harrison, Guy P. *50 Reasons People Give for Believing in a God*. Amherst, NY: Prometheus Books, 2008.

Hartle, James B., and Stephen W. Hawking. "Wave Function of the Universe." *Physical Review* D28 (1983): 2960–75.

Haught, James A. *Holy Horrors: An Illustrated History of Religious Murder and Madness*. Updated ed. Amherst, NY: Prometheus Books, 2002.

———. *Honest Doubt: Essays on Atheism in a Believing Society*. Amherst, NY: Prometheus Books, 2007.

Haught, John F. *God and the New Atheism: A Critical Response to Dawkins, Harris, and Hitchens*. Louisville, KY: Westminster John Knox Press, 2008.

Hauser, Marc, and Peter Singer. "Morality without Religion." http://www.wjh.harvard .edu/~mnkylab/publications/recent/HauserSingerMoralRelig05.pdf (accessed January 24, 2009).

Hawking, Stephen W. *A Brief History of Time: From the Big Bang to Black Holes*. New York: Bantam, 1988.

Hawking, Stephen W., and Roger Penrose. "The Singularities of Gravitational Collapse." *Proceedings of the Royal Society of London* series A, 314 (1970): 529–48.

Hedges, Chris. *American Fascists: The Christian Right and the War on America*. New York: Free Press, 2007.

———. *I Don't Believe in Atheists*. New York: Free Press, 2008.

Hitchcock, S. C. *Disbelief 101: A Young Person's Guide to Atheism*. Tucson, AZ: Sharp Press, 2009.

Hitchens, Christopher. *God Is Not Great: How Religion Poisons Everything*. New York: Twelve, 2007.

———, ed. *The Portable Atheist: Essential Readings for the Nonbeliever*. Philadelphia: Da Capo Press, 2007.

Hume, David. *Dialogues concerning Natural Religion*, edited by J. M. Bell. London: Penguin, 1990.

Ibn Warraq, ed. *The Origins of the Koran: Classic Essays on Islam's Holy Book*. Amherst, NY: Prometheus Books, 1998.

Isaacson, Walter. "Einstein and Faith." http://www.time.com/time/magazine/article/0,9171,1607298,00.html (accessed June 5, 2009).

Jackson, Frank. "Epiphenomenal Qualia." *Philosophical Quarterly* 32 (1982): 127–36.

Jacoby, Susan. *The Age of American Unreason*. New York: Pantheon Books, 2008.

Jaspers, Karl. *The Future of Mankind*. Chicago: University of Chicago Press, 1961.

Johnstone, Brick, and Bert A. Glass. "Support for a Neuropsychological Model of Spirituality." *Zygon* 43, no. 4 (2008): 861–74.

Juergensmeyer, Mark. *Terror in the Mind of God: The Global Rise of Religious Violence*. Berkeley: University of California Press, 2000.

———. *Terror in the Mind of God: The Global Rise of Religious Violence.* 3rd rev. and updated ed. Berkeley: University of California Press, 2003.

Kaufmann, Eric. "Sacralization by Stealth: Demography, Religion, and Politics in Europe." *Jewish Policy Research.* http://www.jpr.org.uk/downloads/Kaufmann%20Paper.pdf (accessed February 25, 2009).

Kirsch, Jonathan. *God against the Gods: The History of the War between Monotheism and Polytheism.* New York: Viking Compass, 2004.

Krakauer, Jon. *Under the Banner of Heaven: A Story of Violent Faith.* 1st Anchor Books ed. New York: Anchor Books, 2004.

Kreeft, Peter. *Faith and Reason: The Philosophy of Religion. A Modern Scholar Audiobook.* Prince Frederick, MD: Recorded Books, LLC, 2005.

Kristol, Irving. *Neoconservatism: The Autobiography of an Idea.* New York: Free Press, 1995.

———. *Two Cheers for Capitalism.* New York: Basic Books, 1978.

Kurtz, Paul. *Eupraxophy: Living without Religion.* Amherst, NY: Prometheus Books, 1989.

———, ed. *Science and Religion: Are They Compatible?* Amherst, NY: Prometheus Books, 2003.

———. *The Transcendental Temptation: A Critique of Religion and the Paranormal.* Amherst, NY: Prometheus Books, 1986.

LaHaye, T. F., and J. B. Jenkins. *Left Behind: A Novel of the Earth's Last Days.* Tyndale House Publishers, 1995.

Lakoff, George, and Mark Johnson. *Philosophy in the Flesh: The Embodied Mind and Its Challenge to Western Thought.* New York: Basic Books, 1999.

Lamont, Corliss. *The Philosophy of Humanism.* 8th ed. Amherst, NY: Humanist Press, 1996.

Larson, Edward J. "Leading Scientists Still Reject God." *Nature* 294, no. 6691 (1998): 313.

Leslie, John, ed. *Modern Cosmology & Philosophy.* Amherst, NY: Prometheus Books, 1998.

Levesque, J., M. Beauregard, and B. Mensour. "Effect of Neurofeedback Training on the Neural Substrates of Selective Attention in Children with Attention-Deficit/Hyperactivity Disorder: A Functional Magnetic Resonance Imaging Study." *Neuroscience Letters* 394 (2006): 216–21.

Libet, B., C. A. Gleason, E. W. Wright, and D. K. Pearl. "Time of Conscious Intention to Act in Relation to Onset of Cerebral Activity (Readiness-Potential): The Unconscious Initiation of a Freely Voluntary Act." *Brain* 196, no. 3 (1983): 623–42.

Lin, Derek, ed. *Tao Te Ching: Annotated & Explained.* Woodstock, VT: SkyLight Paths, 2006.

Lincoln, Bruce. *Holy Terrors: Thinking about Religion after September 11.* Chicago: University of Chicago Press, 2003.

Lind, Michael. "A Tragedy of Errors." *Nation*, February 23, 2004.

Lindsey, Hal, and Carole C. Carlson. *The Late Great Planet Earth*. Grand Rapids, MI: Zondervan, 1970.

Linker, Damon. *The Theocons: Secular America under Siege*. New York: Doubleday, 2006.

Maisel, Eric. *The Atheist's Way: Living Well without Gods*. Novato, CA: New World Library, 2009.

Marcus, Gary F. *Kluge: The Haphazard Construction of the Human Mind*. Boston: Houghton Mifflin, 2008.

Marshall, David. *The Truth behind the New Atheism: Responding to the Emerging Challenges to God and Christianity*. Eugene, OR: Harvest House Publishers, 2007.

Martin, Michael, and Ricki Monnier. *The Impossibility of God*. Amherst, NY: Prometheus Books, 2003.

McGrath, Alister E. *Dawkins' God: Genes, Memes, and the Meaning of Life*. Malden, MA: Blackwell, 2005.

McGrath, Alister E., and Joanna McGrath. *The Dawkins Delusion: Atheist Fundamentalism and the Denial of the Divine*. Downers Grove, IL: InterVarsity Press, 2007.

Melnyk, Andrew. "A Case for Physicalism about the Human Mind." http://www.infidels.org/library/modern/andrew_melnyk/physicalism.html (accessed November 20, 2008).

Miller, Kenneth R. *Only a Theory: Evolution and the Battle for America's Soul*. New York: Viking Penguin, 2008.

Mills, David. *Atheist Universe: Why God Didn't Have a Thing to Do with It*. Philadelphia: Xlibris, 2003.

Mohler, R. Albert. *Atheism Remix: A Christian Confronts the New Atheists*. Wheaton, IL: Crossway Books, 2008.

Mooney, Chris. *The Republican War on Science*. New York: Basic Books, 2005.

Morey, Robert A. *The New Atheism and the Erosion of Freedom*. Phillipsburg, NJ: P&R Publishing, 1994.

Myers, P. Z. "Egnor's Machine Is Uninhabited by Any Ghost." http://scienceblogs.com/pharyngula/2007/06/egnors_machine_is_uninhabited.php (accessed July 12, 2009).

New World Encyclopedia, "Axial Age." http://www.newworldencyclopedia.org/entry/Axial_Age (accessed February 21, 2009).

Norris, Pippa, and Ronald Inglehart. *Sacred and Secular: Religion and Politics Worldwide*. Cambridge Studies in Social Theory, Religion, and Politics. Cambridge: Cambridge University Press, 2004.

Novak, Michael. "Lonely Atheists of the Global Village." *National Review*, March 19, 2007.

———. *No One Sees God: The Dark Night of Atheists and Believers*. New York: Doubleday, 2008.

O'Reilly, Bill. *Culture Warrior.* New York: Broadway Books, 2006.

Paley, William. *Natural Theology; Or, Evidences of the Existence and Attributes of the Deity.* 12th ed. London: Printed for J. Faulder, 1809.

Paul, Gregory. "The Big Religion Questions Finally Solved." *Free Inquiry* 29, no. 1 (2009).

Paul, Gregory, and Phil Zuckerman. "Why the Gods Are Not Winning." http://www.edge.org/3rd_culture/paul07/paul07_index.html (accessed February 8, 2009).

Paulos, John Allen. *Irreligion: A Mathematician Explains Why the Arguments for God Just Don't Add Up.* New York: Hill and Wang, 2008.

Penrose, Roger. *The Emperor's New Mind: Concerning Computers, Minds, and the Laws of Physics.* Oxford: Oxford University Press, 1989.

———. *Shadows of the Mind: A Search for the Missing Science of Consciousness.* Oxford: Oxford University Press, 1994.

Perakh, Mark. *Unintelligent Design.* Amherst, NY: Prometheus Books, 2004.

Petto, Andrew J., and Laurie R. Godfrey. *Scientists Confront Intelligent Design and Creationism.* New York: Norton, 2007.

Pew Research Center. "Many Americans Uneasy with Mix of Religion and Politics." http://pewforum.org/publications/surveys/religion-politics-06.pdf (accessed November 7, 2008).

Phillips, Kevin. *American Dynasty: Aristocracy, Fortune, and the Politics of Deceit in the House of Bush.* New York: Penguin, 2004.

———. *American Theocracy: The Peril and Politics of Radical Religion, Oil, and Borrowed Money in the 21st Century.* New York: Viking, 2006.

Pope John Paul II, "Magisterium Is Concerned with Question of Evolution for It Involves Conception of Man." http://www.cin.org/jp2evolu.html (accessed December 20, 2008).

Price, Huw. *Time's Arrow & Archimedes' Point: New Directions for the Physics of Time.* New York: Oxford University Press, 1996.

Price, Robert M. *Deconstructing Jesus.* Amherst, NY: Prometheus Books, 2000.

———. *The Reason-Driven Life: What Am I Here on Earth For?* Amherst, NY: Prometheus Books, 2006.

Radin, Dean I. *The Conscious Universe: The Scientific Truth of Psychic Phenomena.* New York: HarperEdge, 1997.

Radzinsky, Edvard. *Edvard Radzinsky, Stalin: The First In-Depth Biography Based on Explosive New Documents from Darkness, Tunnels, and Light.* New York: Anchor, 1997.

Rea, Michael. *World without Design: The Ontological Consequences of Naturalism.* Oxford: Clarendon, 2002.

Rees, Martin J. *Just Six Numbers: The Deep Forces That Shape the Universe.* New York: Basic Books, 2000.

Remsburg, John E. *The Christ: A Critical Review and Analysis of the Evidences of His Existence.* Amherst, NY: Prometheus Books, 1994.

Romain, Jonathan. *God, Doubt, and Dawkins: Reform Rabbis Respond to* The God Delusion. London: Movement for Reform Judaism, 2008.

Ross, Hugh. *The Creator and the Cosmos: How the Greatest Scientific Discoveries of the Century Reveal God.* 2nd. expanded ed. Colorado Springs: NavPress, 1995.

Saxton, Alexander. "The Great God Debate and the Future of Faith." *Free Inquiry* 29, no. 1 (2009): 37–43.

Schellenberg, J. L. *Divine Hiddenness and Human Reason.* Ithaca, NY: Cornell University Press, 1993.

Schneider, Peter. *Extragalactic Astronomy and Cosmology: An Introduction.* Berlin: Springer, 2006.

Schwartz, J. M., H. Stapp, and M. Beauregard. "Quantum Physics in Neuroscience and Psychology: A Neurophysical Model of Mind-Brain Interaction." *Philosophical Transactions of the Royal Society* B: Biological Sciences (2005): 1–19.

Searle, John R. *The Rediscovery of the Mind.* Cambridge, MA: MIT Press, 1992.

Seife, Charles. "Cold Numbers Unmake the Quantum Mind." *Science* 287 (2000).

Seth, Anil K., Bernard J. Baara, and David B. Edelman. "Criteria for Consciousness in Humans and Other Animals." *Consciousness and Cognition* 14 (2005): 119–39.

Shermer, Michael. *The Science of Good and Evil: Why People Cheat, Gossip, Care, Share, and Follow the Golden Rule.* New York: Times Books, 2004.

———. "Why People Believe in God: An Empirical Study on a Deep Question." *Humanist* 59, no. 6 (1999): 43–50.

Simon, Stephanie. "Atheists Reach Out—Just Don't Call It Proselytizing." *Wall Street Journal,* November 18, 2008.

Singer, Peter. "God and Suffering, Again." *Free Inquiry* 28, no. 6 (2008): 19–20.

———. *How Are We to Live? Ethics in an Age of Self-Interest.* Oxford: Oxford University Press, 1997.

———. *The President of Good & Evil: The Ethics of George W. Bush.* New York: Dutton, 2004.

Smith, George H. *Atheism: The Case against God.* Amherst, NY: Prometheus Books, 1989.

Stark, Rodney, and Roger Finke. *Acts of Faith: Explaining the Human Side of Religion.* Berkeley: University of California Press, 2000.

Steele, David Ramsay. *Atheism Explained: From Folly to Philosophy.* Chicago: Open Court, 2008.

Stenger, Victor J. *The Comprehensible Cosmos: Where Do the Laws of Physics Come From?* Amherst, NY: Prometheus Books, 2006.

———. "Flew's Flawed Science." *Free Inquiry* 25, no. 2 (2006): 17–18.

———. *God: The Failed Hypothesis—How Science Shows That God Does Not Exist.* Amherst, NY: Prometheus Books, 2007.

———. *Has Science Found God? The Latest Results in the Search for Purpose in the Universe.* Amherst, NY: Prometheus Books, 2003.

———. "A Lack of Evidence." *Physics World* 19, no. 10 (2006): 45–46.

———. "Natural Explanations for the Anthropic Coincidences." *Philo* 3 (2000): 50–67.

———. *Not by Design: The Origin of the Universe.* Amherst, NY: Prometheus Books, 1988.

———. "Onward Science Soldiers." *Skeptical Inquirer* 31, no. 4 (2007): 11–12.

———. *Physics and Psychics: The Search for a World beyond the Senses.* Amherst, NY: Prometheus Books, 1995.

———. *Quantum Gods: Creation, Chaos, and the Search for Cosmic Consciousness.* Amherst, NY: Prometheus Books, 2009.

———. "A Scenario for a Natural Origin of Our Universe." *Philo* 9, no. 2 (2006): 93–102.

———. "Time's Arrow Points Both Ways: The View from Nowhen." *Skeptical Inquirer* 8, no. 4 (2001): 90–95.

———. *Timeless Reality: Symmetry, Simplicity, and Multiple Universes.* Amherst, NY: Prometheus Books, 2000.

———. *The Unconscious Quantum: Metaphysics in Modern Physics and Cosmology.* Amherst, NY: Prometheus Books, 1995.

Stern, Jessica. *Terror in the Name of God: Why Religious Militants Kill.* New York: Ecco, 2003.

Stokes, Douglas M. *The Nature of Mind: Parapsychology and the Role of Consciousness in the Physical World.* Jefferson, NC: McFarland, 1997.

Susskind, Leonard. *Cosmic Landscape: String Theory and the Illusion of Intelligent Design.* New York: Little, Brown, 2005.

Tart, Charles T. "A Psychological Study of Out-of-Body Experiences in a Selected Subject." *Journal of the American Society for Psychical Research* 62 (1968): 3.

Tegmark, Max. "The Importance of Quantum Decoherence in Brain Processes." *Physical Review E* 61 (1999): 4194–206.

Trottier, Justin. "Atheists: Necessarily Arrogant?" *Free Inquiry* 28, no. 6 (2008): 51–52.

Van Hejenport, Jean, ed. *A Source Book in Mathematical Logic, 1879–1931.* 1967.

Wade, Nicholas. *Before the Dawn: Recovering the Lost History of Our Ancestors.* New York: Penguin, 2006.

Ward, Keith. *In Defence of the Soul.* Oxford: Oneworld, 1998.

Warren, Richard. *The Purpose-Driven Life: What on Earth Am I Here For?* Grand Rapids, MI: Zondervan, 2002.

Wells, George Albert. *The Jesus Myth.* Chicago, IL: Open Court, 1999.

White, Andrew Dickson. *A History of the Warfare of Science with Theology in Christendom.* New York: D. Appleton, 1896.

Wielenberg, Erik J. *Value and Virtue in a Godless Universe.* Cambridge: Cambridge University Press, 2005.

Williams, Thomas D. *Greater Than You Think: A Theologian Answers the Atheists about God.* New York: Faith Words, 2008.

Wilson, David Sloan. *Darwin's Cathedral: Evolution, Religion, and the Nature of Society.* Chicago: University of Chicago Press, 2002.

Woerlee, G. M. "Darkness, Tunnels, and Light." *Skeptical Inquirer* 28, no. 3 (2004).

———. *Mortal Minds: The Biology of Near-Death Experiences.* Amherst, NY: Prometheus Books, 2005.

Young, Matt, and Taner Edis. *Why Intelligent Design Fails: A Scientific Critique of the New Creationism.* New Brunswick, NJ: Rutgers University Press, 2004.

Zacharias, Ravi K. *The End of Reason: A Response to the New Atheists.* Grand Rapids, MI: Zondervan, 2008.

Zakaria, Fareed. *The Future of Freedom: Illiberal Democracy at Home and Abroad.* New York: Norton, 2003.

Zindler, Frank R. *The Jesus the Jews Never Knew: Sepher Toldoth Yeshu and the Quest of the Historical Jesus in Jewish Sources.* Cranford, NJ: American Atheist Press, 2003.

Zuckerman, Phil. *Society without God: What the Least Religious Nations Can Tell Us about Contentment.* New York: New York University Press, 2008.

Zusne, Leonard, and Warren Jones. *Anomalistic Psychology: A Study of Extraordinary Phenomena of Behavior and Experience.* Hillside, NJ: Lawrence Erlbaum Associates, 1982.

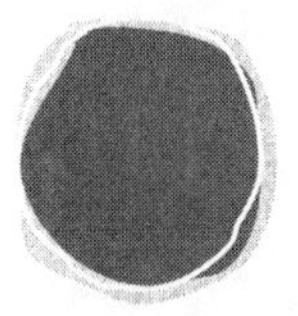

ACKNOWLEDGMENTS

As in my other books, I have relied heavily on feedback from the e-mail discussion list avoid-L@hawaii.edu. I am especially grateful to Bob Zannelli for taking over the burden of managing this list for me as well as for his many helpful comments and suggestions on this manuscript. Thanks also to Lawrence Crowell, Yonatan Fishman, John Kole, Don McGee, Brent Meeker, Anne O'Reilly, Christopher Savage, Kerry Regier, and Taner Edis. And, as always, I could not function without the loving support of my wife, Phylliss, daughter, Noelle Green, son, Andy, son-in-law, Joe Green, and daughter-in-law, Helenna Nakama. And bringing the joy to my life that makes it all worthwhile are my grandchildren Katie Stenger, Lucy Green, Zoe Stenger, and Joey Green.

Finally, I need to acknowledge editor in chief, Steven L. Mitchell, and his excellent editorial and production staff, notably Joe Gramlich, Chris Kramer, and Grace Zilsberger, for their hard work and professionalism.

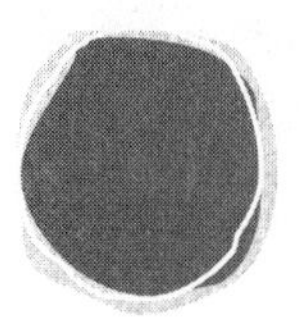

INDEX

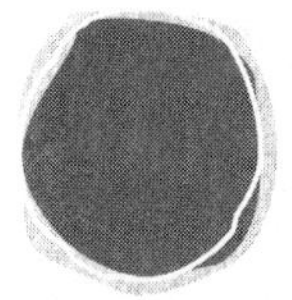

ABOUT THE AUTHOR

Victor J. Stenger grew up in a Catholic working-class neighborhood in Bayonne, New Jersey. His father was a Lithuanian immigrant, his mother the daughter of Hungarian immigrants. He attended public schools and received a bachelor's of science degree in electrical engineering from Newark College of Engineering (now New Jersey Institute of Technology) in 1956. While at NCE, he was editor of the student newspaper and received several journalism awards.

Moving to Los Angeles on a Hughes Aircraft Company fellowship, Dr. Stenger received a master's of science degree in physics from UCLA in 1959 and a PhD in physics in 1963. He then took a position on the faculty of the University of Hawaii, retiring to Colorado in 2000. He currently is emeritus professor of physics at the University of Hawaii and adjunct professor of philosophy at the University of Colorado. Dr. Stenger is a fellow of the Committee for Skeptical Inquiry and a research fellow of the Center for Inquiry. Dr. Stenger has also held visiting positions on the faculties of the University of Heidelberg in Germany, Oxford in England (twice), and has been a visiting researcher at Rutherford Laboratory in England, the National Nuclear Physics Laboratory in Frascati, Italy, and the University of Florence in Italy.

His research career spanned the period of great progress in elementary particle physics that ultimately led to the current *standard model.* He participated in experiments that helped establish the properties of strange particles, quarks, gluons, and neutrinos. He also helped pioneer the emerging fields of very high-energy gamma-ray and neutrino astronomy. In his last project before retiring, Dr. Stenger collaborated on the underground experiment in Japan that in 1998 showed for the first time that the neutrino has mass. The Japanese leader of this experiment shared the 2002 Nobel Prize for this work.

Victor Stenger has had a parallel career as an author of critically well-received popular-level books that interface between physics and cosmology and philosophy, religion, and pseudoscience. These include: *Not by Design: The Origin of the Universe* (1988); *Physics and Psychics: The Search for a World beyond the Senses* (1990); *The Unconscious Quantum: Metaphysics in Modern Physics and Cosmology* (1995); *Timeless Reality: Symmetry, Simplicity, and Multiple Universes* (2000); *Has Science Found God? The Latest Results in the Search for Purpose in the Universe* (2003); *The Comprehensible Cosmos: Where Do the Laws of Physics Come From?* (2006); *God: The Failed Hypothesis—How Science Shows That God Does Not Exist* (2007); and *Quantum Gods: Creation, Chaos, and the Search for Cosmic Consciousness* (2009). *God: The Failed Hypothesis* last made the *New York Times* Best Seller List in March 2007.

Dr. Stenger and his wife, Phylliss, have been happily married since 1962 and have two children and four grandchildren. They now live in Lafayette, Colorado. They attribute their long lives to the response of evolution to the human need for babysitters, a task they joyfully perform. Phylliss and Vic are avid doubles tennis players and generally enjoy the outdoor life in Colorado, and they travel the world as often as they can.

Dr. Stenger maintains a popular Web site (a thousand hits per month), where much of his writing can be found, at http://www.colorado.edu/philosophy/vstenger.